Viviana Palumberi

Chondrocyte growth dynamics and spatial pattern formation

Viviana Palumberi

Chondrocyte growth dynamics and spatial pattern formation

Mathematical modeling

Südwestdeutscher Verlag für Hochschulschriften

Impressum/Imprint (nur für Deutschland/only for Germany)
Bibliografische Information der Deutschen Nationalbibliothek: Die Deutsche Nationalbibliothek verzeichnet diese Publikation in der Deutschen Nationalbibliografie; detaillierte bibliografische Daten sind im Internet über http://dnb.d-nb.de abrufbar.

Alle in diesem Buch genannten Marken und Produktnamen unterliegen warenzeichen-, marken- oder patentrechtlichem Schutz bzw. sind Warenzeichen oder eingetragene Warenzeichen der jeweiligen Inhaber. Die Wiedergabe von Marken, Produktnamen, Gebrauchsnamen, Handelsnamen, Warenbezeichnungen u.s.w. in diesem Werk berechtigt auch ohne besondere Kennzeichnung nicht zu der Annahme, dass solche Namen im Sinne der Warenzeichen- und Markenschutzgesetzgebung als frei zu betrachten wären und daher von jedermann benutzt werden dürften.

Coverbild: www.ingimage.com

Verlag: Südwestdeutscher Verlag für Hochschulschriften GmbH & Co. KG
Heinrich-Böcking-Str. 6-8, 66121 Saarbrücken, Deutschland
Telefon +49 681 37 20 271-1, Telefax +49 681 37 20 271-0
Email: info@svh-verlag.de

Approved by: Basel, Diss., 2009

Herstellung in Deutschland:
Schaltungsdienst Lange o.H.G., Berlin
Books on Demand GmbH, Norderstedt
Reha GmbH, Saarbrücken
Amazon Distribution GmbH, Leipzig
ISBN: 978-3-8381-3164-1

Imprint (only for USA, GB)
Bibliographic information published by the Deutsche Nationalbibliothek: The Deutsche Nationalbibliothek lists this publication in the Deutsche Nationalbibliografie; detailed bibliographic data are available in the Internet at http://dnb.d-nb.de.

Any brand names and product names mentioned in this book are subject to trademark, brand or patent protection and are trademarks or registered trademarks of their respective holders. The use of brand names, product names, common names, trade names, product descriptions etc. even without a particular marking in this works is in no way to be construed to mean that such names may be regarded as unrestricted in respect of trademark and brand protection legislation and could thus be used by anyone.

Cover image: www.ingimage.com

Publisher: Südwestdeutscher Verlag für Hochschulschriften GmbH & Co. KG
Heinrich-Böcking-Str. 6-8, 66121 Saarbrücken, Germany
Phone +49 681 37 20 271-1, Fax +49 681 37 20 271-0
Email: info@svh-verlag.de

Printed in the U.S.A.
Printed in the U.K. by (see last page)
ISBN: 978-3-8381-3164-1

Copyright © 2012 by the author and Südwestdeutscher Verlag für Hochschulschriften GmbH & Co. KG and licensors
All rights reserved. Saarbrücken 2012

To Dalia

"... a model must be wrong, in some respects
– else it would be the thing itself.
The trick is to see ... where it is right."
Henry Bent

Acknowledgments

This thesis was written at the Mathematical Institute, University of Basel, Switzerland. It was supported by the University of Basel and for one year by the Marie Heim-Vögtlin-Programm of the Swiss National Foundation under the project PMCD22-118601.

First of all, I would like to thank Prof. Dr. Marcus J. Grote for all the hours we spent together discussing and overcoming the difficulties and problems we encountered during this study.

Then, thanks to Dr. Barbara Wagner (WIAS, Berlin) not only for being my co-referee, but also for all the helpful suggestions she gave me during my doctoral thesis and to Prof. Dr. Ivan Martin (ICFS University Hospital, Basel) for his practical advice always useful for making steps forward.

I am also much indebted to Prof. Dr. Assyr Abdulle (EPFL, Lausanne) who was always interested in my work during his days in Basel and introduced me to the Chebyshev methods. Moreover, I would like to thank Prof. Dr. Ben Schweizer, Dr. Marco Veneroni and Dr. Michael Lenzinger (Technische Universität Dortmund) for their help in the analytical study of our model as well as Prof. Dr. Thomas Vetter (Computer Science Department, Basel) for his crucial advice about the use of Gabor filters.

To my ex-boss Peter Maria Engeli (Consultant, Vectoris AG) I am really grateful for having introduced me not only to the C-programming, but also to the swiss language and life. Moreover, I don't want to forget Dr. Mischa Reinhardt (Novartis, Basel) who followed me as mentor during the WIN-Program and in particular gave me precious tips about giving a talk in public.

To my colleagues at the Mathematical Institute in Basel go my sincere thanks, in particular to Dr. Anna Schneebeli (now Credit Suisse) for her friendship and support from the beginning to the end, to Dr. Teodora Mitkova and Dr. David Cohen for their patience in reading and helping me improving the draft and to Christian Stohrer for his kindness and constant help in the preparation of the tutorial which allowed me to dedicate more time to this work.

I thank also all my friends, both from Italy and Switzerland, for being always present and my parents and brothers who are constantly close to me although more than five hundred kilometers keep us apart.

Finally, my special thanks go to my husband Dr. Andrea Barbero for his constant love and support both in the work and real life.

Contents

1. Introduction   **1**
   1.1. Historical overview 1
   1.2. Pattern formation in cell culture 2
   1.3. Study of chondrocyte culture 3

2. Experimental and mathematical study of the influence of growth factors on the growth kinetics of chondrocytes   **5**
   2.1. Introduction 5
   2.2. Material and methods 6
      2.2.1. Cell culture 6
      2.2.2. Mathematical model 8
      2.2.3. Statistical analysis 10
   2.3. Results 10
      2.3.1. Growth curves 10
      2.3.2. Microcolony tests 10
      2.3.3. Numerical simulations 13
      2.3.4. Dierence in the growth kinetic between AHAC at dier ent passages in culture . . 14
      2.3.5. Discussion 17

3. Dynamic Formation of Oriented Patches in Chondrocyte Cell Cultures   **21**
   3.1. Introduction 21
   3.2. Biological background 23
      3.2.1. The impact of growth factors 23
      3.2.2. Cell culture: isolation and expansion 23
      3.2.3. Image analysis of alignment 25
   3.3. Mathematical Model 26
      3.3.1. Formulation 26
      3.3.2. Numerical Methods 29
   3.4. Stability 31
   3.5. Comparison of simulations with experiments 33
      3.5.1. Parameter values 33
      3.5.2. Numerical simulations 34
   3.6. Concluding remarks 34

4. Analysis of the spatio-angular model   **40**
   4.1. Existence of a weak solution 40
      4.1.1. Maximum principle and mass control 42
      4.1.2. Galerkin approximations 43
      4.1.3. A-priori estimates 44
      4.1.4. Existence 46
   4.2. Linear stability analysis 47
      4.2.1. Linearization of the original model 49

	4.2.2. Analysis of the instability condition	49
	4.2.3. Linearization of the extended model	54
4.3.	Numerical methods	57
	4.3.1. Integration in space and angle	57
	4.3.2. Integration in time	60
	4.3.3. Numerical simulations	64

5. Conclusions   70

A. Logistic equation   72
    A.1. Classical logistic equation 72
    A.2. Delay logistic equation 72
         A.2.1. Delay in the linear term 73
         A.2.2. Delay in the quadratic term 73

B. Spatio-angular model, calculations and inequalities   76
    B.1. Governing Equation 76
    B.2. Normalization of the kernels 77
    B.3. Inequalities 77

C. Spatial diffusion   79
    C.1. Random walks 79
    C.2. Experiments 80

1. Introduction

> "La filosofia è scritta in questo grandissimo libro che continuamente ci sta aperto innanzi a gli occhi (l'Universo), ma non si può intendere se prima non s'impara a intender la lingua e conoscer i caratteri, ne' quali è scritto. Egli è scritto in lingua matematica, e i caratteri son triangoli, cerchi ed altre figure geometriche, senza i quali mezzi è impossibile a intenderne umanamente parola; senza questi è un aggirarsi vanamente in un oscuro labirinto". [1] Galileo Galilei(1564-1642)

1.1. Historical overview

Galileo was the first important scientist who applied mathematics to physics; he was followed by René Descartes (1596-1650), who introduced the Cartesian axes and the summit was reached with Newton (1642-1727), who created, at the same time as Leibniz (1632-1677), the infinitesimal calculation. The genius of these scientists was that they understood how the complexity of physical phenomena could be brought to a set of mathematical relationships. However, the revolution that they made did not concern the life sciences like biology. Three centuries passed before a pioneer of mathematics, Vito Volterra (1860-1940), put the first stone of the new discipline, the biomathematics. His son-in-law Umberto D'Ancona was a biologist and has gathered detailed fishery statistics between 1905 and 1923 at Venice, Trieste and Fiume. Studying them he had noticed an increase in population of predators in the Adriatic Sea during World War I as compared with the preceding and following periods, as well as the decrease in their prey. The study related to the ongoing debate about the need for fishery regulation.

The main point was that after fishing resumed at the end of the War, no global increase occurred in the fish market, but only a relative increase of some species and decrease of others. According to D'Ancona, the suspension of fishery during the war displaced the biological equilibrium in the Adriatic Sea in favor of the voracious species. He discussed the problem with his father-in-law Vito Volterra who gave a theoretical explanation of these variations in the relative numbers of a biological association introducing a system of two ordinary dierential equations . Volterra published his first results in 1926 [49].

Vito Volterra played a decisive and widely acknowledged role in the modern developments of mathematical biology. In the period prior to the Second World War quite a lot of seminal work towards a systematic and organic development of mathematical research in biology was produced. One of the most important contribute had been given by the statistician Ronald A. Fisher (1890-1962) who introduced a whole set of mathematical tools to deal with problems in population genetics. What distinguishes Fisher from Volterra is that the first one used extensively probabilistic techniques whereas the second one held a deterministic point of view expressed in dierential equa tions, as we also did in this work.

Some preliminary remarks are necessary. The merely instrumental applications of mathematics to biology, that is, the use of elementary computations, must be distinguished from eorts towards a conceptual application of mathematics to biology, that is, the building of a genuine methodology of biomathematical research. As part of his general scientific reductionist program Volterra aimed his biomathematical

---

[1] The philosophy is written in this great book that is continually open in front of our eyes (the Universe), but we can not understand it until we do not learn to understand the language and the characters in which it is written. It is written in the mathematical language and the characters are triangles, circles and other geometric figures, without these means is humanly impossible to understand a word; without them it is a vain wandering in a dark labyrinth.

research to transfer the conceptual apparatus of mechanics to biology. In his early studies he applied dierential equations and integro-dierential equations to build a rational mechanics of biological associations. Many biologists questioned the legitimacy of applying mathematical concepts and methods in biology. They felt that biology, as a natural science concerned with living beings, could not follow the blind mathematical laws of physics or submit to the simplification requirements of a mathematical formulation. An important point in all these discussions was the fit of biomathematical equations and laws to experience. Practical problems in agriculture as well as fishery motivated an indistinct interest in quantitative methods in biology. However, many biologists did not trust the possible eective results of mathematical research. On the other hand, also the classical science was undergoing an important transformation, trying to give a description of natural phenomena through mathematical laws that were subjected to comparison with experimental data. Volterra wanted to extend this schema to biology [27].

However, mathematics can not be applied to biology in the same way as to classical physics; the role of a mathematical model in biology is dierent. A good model s hould help to understand the behavior of biological systems with the aim of intervening in a more e ective way, should be able to find the links among the information we gather from the real world, in order to anticipate the evolution of the behavior for example of a cell or a group of cells. The mathematics is an instrument of the human mind to intervene on the nature, it is a human creation to understand the world in order to operate on it. Mathematics is not only the exact language of the science and the nature, but also an instrument to better understand and intervene on them, to build things for the satisfy of the humanity's necessities. After Volterra, and in particular in the last twenty years, the growth of mathematical biology and the diversity of applications has been astonishing. Mathematical modeling is being applied in every major discipline in the biomedical science.

1.2. Pattern formation in cell culture

Here, we developed a mathematical model to better understand the behavior of some particular cells. As Volterra did, we opted for a deterministic approach. Another powerful instrument to understand the living phenomena is the stochastic approach, but it would have required a detailed study of the single cell behavior, whereas the experimental data we could gather gave us only a macroscopic description of the cell culture. Deterministic models and in particular reaction-diusion systems have been widely used in the study of biological phenomena, as wound healing, patterns formation or tumors growth. The study of the pattern generation is for example very important in embryology where the mathematical models provide with possible scenarios as how pattern is laid down and how the embryonic form might be created. In [37] dierent models are proposed, such as tha t for the embryonic fingerprint formation which can be compared to the pattern formation created by cells cultured on a two-dimensional plate.

Regarding the generation of patterns in cell culture we refer to the important work of Elsdale [17] where fibroblast cultures were analyzed to investigate how densely packed cells organize. In particular normal human lung fibroblasts were cultured and cell movement and patterning were studied with time-lapse cinemicrography. The cells spread randomly and eventually stabilized forming a dense patchwork of arrays of fibroblasts as confluences were approached. As a result, the confluent culture formed a patchwork of numerous parallel arrays where the cells were no longer free but constrained in their movement along some lines. The arrays merged at confluence where the cells in two adjacent arrays shared the same orientation to within a small angle. Experiments in [19] complement those of Elsdale and indicate that when cells come into contact with each other at a small angle, only a small portion of the filopodial protrusions are inhibited and neighboring cells glide along and adhere at each other. At large angles of contact cells may crawl over each other or move away from each other. The angle of contact that produces this feature was suggested for dierent tissues.

A mathematical model was introduced in [16] to prove that the pattern formation can be caused by the mere interactions of individual cells, although it is a population phenomena. Until then the formation of structures was only attributed to other mechanisms as chemical gradients (chemotaxis) or mechanical stresses. In this regard we refer in particular to [38] where a mathematical analysis was proposed to understand how these mechanisms conspire to generate organized spatial aggregations. In [16] indeed the authors showed that the self-organization of cells can actually be explained from contact-responses of the cells alone. Their integro-dierential equations cons idered the distribution of the cells as a variable of the time and the angle of orientation. They presented two equations, one for cells that are bounded and one for free cells, but in [34] a more physical approach was considered leading to a single equation. Furthermore in [35] the model was extended to also take into account the spatial distribution of the cells. We could not find in the literature any similar works applied to the cells considered in this dissertation. Because of the similarities between these cells and the fibroblasts we decided to start from this last model.

1.3. Study of chondrocyte culture

We worked in tight collaboration with the Tissue Engineering Group (TEG) at University Hospital in Basel. Cartilage tissue engineering is a novel and promising approach to repair articular cartilage defects. This procedure requires that cartilage cells (chondrocytes) are isolated from a small biopsy and expanded in vitro, generally on two-dimensional culture plates (monolayer), to augment their original number. Post expanded cells are then cultured on specific biosynthetic materials and grafted in the cartilage defects. One of the challenges that arise in this procedure is that the chondrocytes undergo only a limited number of divisions in vitro. A possible way to overcome this limit consists in the supplantation of specific bioactive molecules (growth factors) during the culture of chondrocytes. In this regard, the TEG developed an innovative growth factors combination (TGFβ1, FGF-2, and PDGF BB) that accelerates the growth of the chondrocytes on a monolayer [5], as it can be seen in Fig. (1.1). To investigate how these growth factors influence the cell expansion we were asked to seek an appropriate mathematical model.

In a first step, we developed a model combining time-lag (delay) and logistic equations to capture the kinetic parameters and to enable the description of the complete growth process of the cell culture. The results have been published in [6] and are presented here in chapter 2. However, this model only describes how the number of cells changes in time, without considering the spatial evolution of the cells on a two-dimensional substrate. In previous experiments we observed that chondrocytes cultured with growth factors change not only their shapes, but also their main characteristics, being then very similar to fibroblasts [7]. This suggested that we start from the model developed in [35] which, however, does not consider the cell duplication. We extended this model in an innovative way, adding a logistic terms to follow the cell dynamics during the entire culture time. In particular, we used this model to analyze the formation of patterns at confluence. Indeed it was observed in experiments that when the density of the cells reaches a critical level there is a spontaneous tendency to align along some common axis of orientation. The selection of a preferred axis of orientation can be explained by the fact that the uniform steady state (one in which cells are uniformly distributed in orientation and space) could be unstable under particular conditions. We used linear stability theory to test for the presence of such instability. Indeed, bifurcations can lead to loss of stability of a uniform steady state in favor of patterned states, where cells are aligned in parallel arrays or aggregated in clusters. We remark that we always tried not to loose the link with the biological context by discussing constantly our results with the TEG. In particular, for the comparison them with biological experiments it was essential to use sophisticated image analysis tools which also permit to analyze the orientation of the cells.

In summary, in chapter 2 we present a model to approximate some important kinetic parameters which we then used in the spatial model introduced in chapter 3. Here, we investigate the spatial characteristics of the cells and we develop mathematical tools to calculate the number of patterns arising at confluence as well as their size. In chapter 4 we study in details the spatial model from dierent points of view. We prove the existence under specific conditions of a weak solution through Galerkin approximations and a-priori estimates. Then we perform a linear stability analysis to have information about parameter regimes that give arise to a formation of patterns. Integrating the model with a combination of Chebyshev methods, finite dierences and trapezoidal quadrature we an alyze the behavior of the model for dierent parameter regimes. In the appendix we report calculations, proofs and experiments as supplements to the previous chapters.

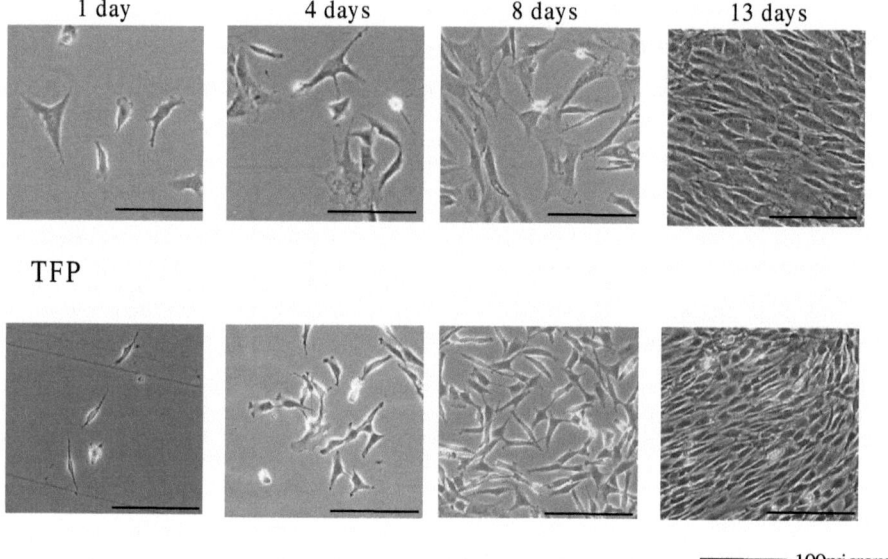

Figure 1.1.: *Representative pictures at different times of culture of human articular chondrocytes expanded without (CTR) or with growth factors (TFP).*

One of the main problems the TEG is confronting consists in the variability of the behavior of chondrocytes isolated from dierent donors. In a study performe d to investigate age related changes in proliferation and post-expansion tissue-forming capacity [5] an extreme variability in these properties was unexpectedly observed among chondrocytes derived from donors within the same age range. In this regard, the model we present could help biologists either in defining conditions that improve chondrocyte properties or in identifying donor cells that have adequate characteristics for clinical application.

2. Experimental and mathematical study of the influence of growth factors on the growth kinetics of chondrocytes

The content of this chapter has been published in [6].

This study aimed at determining how kinetic parameters of adult human articular chondrocytes (AHAC) growth are modulated by the growth factor combination TGFβ1, FGF-2, and PDGF BB (TFP), recently shown to stimulate AHAC proliferation. AHAC, isolated from cartilage biopsies of three individuals, were cultured in medium without (CTR) or with TFP. For growth curves, AHAC were seeded at 1000 cells/cm^2 and cultured for 12 days, with cell numbers measured fluorimetrically in the same wells every 12 hours. For microcolony tests, AHAC were seeded at 2.5 cells/cm^2 and cultured for 6 days, with cell numbers determined for each microcolony by phase contrast microscopy every 8 hours. A mathematical model combining delay and logistic equations was developed to capture the growth kinetic parameters and to enable the description of the complete growth process of the cell culture. As compared to CTR medium, the presence of TFP increased the number of cells/well starting from the fifth day of culture, and a 4-fold larger cell number was reached at confluence. For single microcolonies, TFP reduced the time for the first cell division by 26.6%, the time for subsequent cell divisions (generation time) by 16.8%, and the percentage of quiescent cells by 42.5%. The mathematical model fitted well the experimental data of the growth kinetic. Finally, using both microcolony tests and the mathematical model, we determined that prolonged cell expansion induces an enrichment of AHAC with shorter first division time, but not of those with shorter generation time.

2.1. Introduction

In several cell therapy applications, the use of cytokines during cell expansion has been proposed as a promising method to increase the number of cells that can be obtained starting from a small biopsy, particularly for cell types with limited proliferative capacity [8, 9, 39, 43, 45]. Depending on the cell system under investigation, the cytokine-induced increase in cell number may underlay a variety of causes, such as a shorter time to start the first cell division, a shorter cell division time, a lower percentage of quiescent cells, and/or a larger density of cells reached at confluence. For example, in [15] is showed that IL-2 influences T-cell proliferation by increasing the proportion of cells that enter the first division and by reducing the average division time, but not by altering the time at which the cells enter the first division. In another study, [14] it was observed that specific growth factors (i.e., FGF-2, EGF, SCF or IGF-1) induced proliferation of muscle-derived stem cell by recruitment into the cell cycle in case of freshly isolated cells, or by reducing the length of the cell cycle in case of an expanded clone. Taking together these studies demonstrate the importance of monitoring several parameters of cell growth following stimulation with growth factors. Quantifying the appropriate kinetic parameters may also be relevant to investigate whether the eect of cytokin es is related to a possible selection of certain subpopulations, and to develop realistic mathematical models characterizing and predicting cell growth. The most simple and frequently used mathematical models apply equations of exponential growth to estimate the population doubling time. The key assumption of these models is that all the cells in culture divide at the same time; therefore, the estimated doubling time reflects a macroscopic feature of the cell culture, which does not take into account the properties of individual cells. Non-exponential time-lag models have been shown to overcome these limitations and provide more realistic estimation of several parameters of cell growth kinetics [4, 42] . However, to our knowledge these models have not yet been combined with logistic equations to model contact inhibition upon cell confluence. Recently we reported that the number of adult human articular chondrocytes (AHAC) obtained following monolayer culture is markedly increased by the use of TGFβ1, FGF-2, or PDGF-BB [28], especially when used in combination (TFP) [5]. Here we aimed at determining how kinetic parameters of AHAC growth are modulated by TFP. In particular, we first used microcolony assay [41] to estimate the following kinetic parameters

- time of first cell division t_{cd},
- cell division time of single cells, generation time G_T,
- percentage of quiescent cells Q_c,
- fraction of cells that divide per generation time $F_c(T)$.

We then developed a mathematical model combining time-lag (delay) and logistic equations to capture the kinetic parameters and to enable the description of the complete growth process of the cell culture. Finally, using the experimental and mathematical methods, we assessed the growth kinetic parameters of AHAC from the same donor at dierent passages in culture, to d etermine whether prolonged expansion in the presence or absence of TFP induces an enrichment in the fraction of the fastest proliferating cells.

2.2. Material and methods

2.2.1. Cell culture

Cell isolation and expansion

Full-thickness human articular cartilage samples were collected from the femoral lateral condyle of three individuals (patient A: 52 years, patient B: 50 years, patient C: 52 years), with no history and no radiographic signs of joint disease, after informed consent and in accordance with the local Ethical Commission. Human adult articular chondrocytes (AHAC) were isolated using 0.15% type II collagenase for 22 hours and resuspended in Dulbecco's modified Eagle's medium (DMEM) containing 10% foetal bovine serum, 4.5 mg/ml D-Glucose, 0.1 mM nonessential amino acids, 1 mM sodium pyruvate, 100 mM HEPES buer, 100 U/ml penicillin, 100 5g/ml streptomycin, a nd 0.29 mg/ml L-glutamine (complete medium). The isolated AHAC were counted using trypan blue, plated in tissue culture flasks at a density of 104 cells/cm^2 and cultured in complete medium, either without growth factors (control medium, CTR) or with the addition of 1 ng/ml of Transforming Growth Factor-b1 (TGF-β1), 5 ng/ml of Fibroblast Growth Factor-2 (FGF-2) and 10 ng/ml of Platelet-Derived Growth Factor-BB (PDGF-BB) (growth factor medium, TFP) in a humidified 370C/5% CO2 incubator. When cells were approximately 80% confluent, first passage (P1) cells were rinsed with phosphate buered saline, detached using 0.05% trypsin/0.53mM EDTA and frozen in complete medium containing 10AHAC were then used for the kinetic studies described below (i.e., growth curves and microcolony tests) either immediately after thawing (passage 1 cells, P1) or following an additional expansion for 2 weeks (passage 2 cells, P2).

Growth curves

AHAC were seeded in 6 well-plates in CTR or TFP medium at a density of 1000 cells/cm^2 and placed in a humidified 370C/5% CO2 incubator. Cell proliferation was assessed during 12 days' culture by

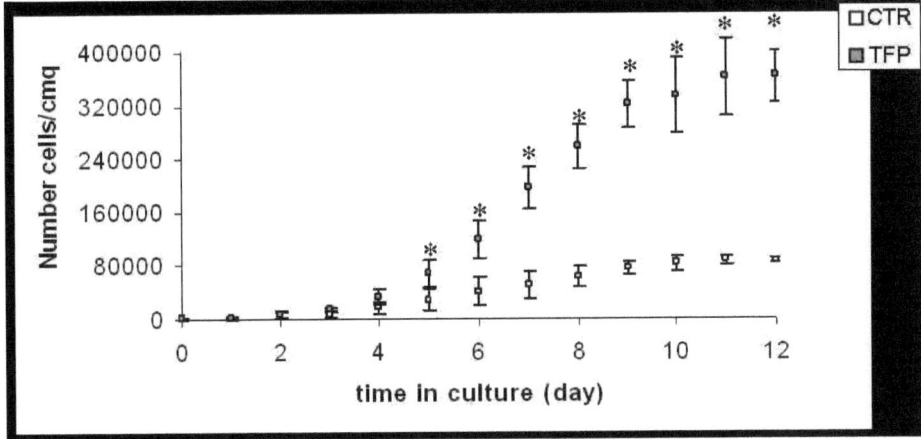

Figure 2.1.: *Growth curves of AHAC expanded in CTR and TFP medium; cell number was measured fluorimetrically every 12 hours and normalized to the dish surface area; values are the mean ± SD of cells from three donors. (*)$P < 0.05$ from AHAC expanded in CTR.*

repeated measures of cell numbers in the same wells (N=6 for each experiment) using alamarBlue^{TM} (a component transformed by living cells from the oxidized non-fluorescent state to the reduced fluorescent state). Briefly, at 12 hours intervals, culture medium was replaced with fresh medium containing 10% alamarBlue^{TM} solution (Serotec Ltd, Düsseldorf, Germany). After four hours, fluorescence intensity was measured (excitation: 560 nm; emission: 590 nm) and converted to cell numbers using a standard curve, generated in preliminary experiments by haemocytometer-based counting of trypsinized cells immediately after alamarBlue^{TM} assay. Morphological features of AHAC cultures in CTR and TFP media were monitored by phase contrast microscopy.

Microcolony tests

Cell culture flasks (150 cm^2) were prepared by drawing a grid below the culture surface (3mm-spaced horizontal and vertical lines). AHAC were seeded in the flasks in CTR or TFP medium at a density of 2.5 cells/cm^2. The use of such a low cell density allowed observation of microcolonies derived from single cells, whereas the use of the grid allowed to track the same microcolonies at dierent times. The number of cells in each microcolony was manually counted using phase contrast microscopy at 8 hour intervals for 6 days. The collected data were used to calculate

- the time of first cell division t_{cd}, as the time (approximated as a multiple of 8 hours) required by each seeded cell to reach the stage of a 2-cell microcolony;

- the generation time g_T, as the time (approximated as a multiple of 8 hours) required by each 2-cell microcolony to reach the stage of a 4-cell microcolony (in preliminary experiments, we found that g_T was virtually identical to the time required by each 4-cell microcolony to reach the stage of a 8-cell microcolony; therefore, g_T can be considered as the cell division time following the first cell division);

- the percentage of quiescent cells Q_c, defined as those which did not reach the stage of 2-cell microcolonies during the entire observation time;
- the fraction of cells that divide per generation period $f_c(T)$, defined as

$$f_c(T) = \frac{\sum_{i=1}^{N-1} DF_i}{\frac{T}{g_T}} \qquad (2.1)$$

where DF_i (dividing fraction in the i-th interval) is the ratio between the new cells that appear in the i-th interval and the cells at the previous interval, N is the number of observations, and T is the total observation time (144 hours) [41].

2.2.2. Mathematical model

Description of the model

An exponential model ($dN/dt = \rho N(t)$) assumes that all cells divide instantaneously, so that the growth rate at time t is proportional to $N(t)$, the number of cells at time t. Based on preliminary experiments we found that this assumption is not correct for AHAC, since the first cell division is not instantaneous. This prompted for the use of delay dierential equations (DD Es) where the growth rate at time $t \geq G_T$ is set proportional to the cell number at some previous time $(t - G_T)$ [4], where G_T indicates the average generation time of the cell population. In appendix A the dierent logistic equations are analyzed in details. Here, we began with investigating the following simple delay model

$$\frac{dN}{dt} = \rho N(t - G_T), \quad t > 0, \quad N(0) = N_0 \qquad (2.2)$$
$$N(t) = \psi(t), \quad -G_T \leq t < 0.$$

where ρ is the cell proliferation rate and, assuming that there is no relevant cell death, corresponds to the number of ospring per parent cell in the population per time unit. The experiment is assumed to start at time $t = 0$. Over the first interval $[0, G_T)$, the rate of growth depends on some previous fictitious cell number. Hence, it is necessary to specify a function $\psi(t)$ over $[-G_T, 0)$ that defines the rate at which new cells appear over $[0, G_T)$. However, one should not interpret $\psi(t)$ as the number of cells $N(t)$ for negative t in $[-G_T, 0)$, but rather $\rho \psi(t - G_T)$ as the rate of the cell growth for t in $[0, G_T)$. If the growth is synchronous and the cells divide around some specific time, $\psi(t)$ should be a Gaussian centered about that time, but if the growth is asynchronous, $\psi(t)$ should be a constant. In either case, the function is normalized by assuming that the number of cells duplicates over the first interval $[0, G_T)$. Integrating (2.2) over $[0, G_T)$, where $N(t - G_T) = \psi(t - G_T)$ and imposing this normalization on $\psi(t)$, we find the condition which the integral of $\psi(t)$ must satisfy. Since our data presented an asynchronous behavior, we chose $\psi(t)$ constant, with $\psi(t) = N_0/(G_T \rho)$. Beyond the seventh day the cell population encounters the physical limitation of the well size. To model the growth kinetics of the cells throughout the entire experiment, that is until confluence, we introduced a logistic delay equation

$$\frac{dN}{dt} = \rho N(t - G_T)\left(1 - \frac{N(t)}{K}\right), \quad t > 0, \quad N(0) = N_0 \qquad (2.3)$$
$$N(t) = \psi(t), \quad -G_T \leq t < 0.$$

Here the growth rate is no longer the single parameter ρ but is given by $\rho(1 - N(t)/K)$, which decreases to zero as the number of cells $N(t)$ tends to the constant K. The parameter K defines the carrying capacity of the environment. In our experiments, K is determined by the space available to the cells, i.e. the well

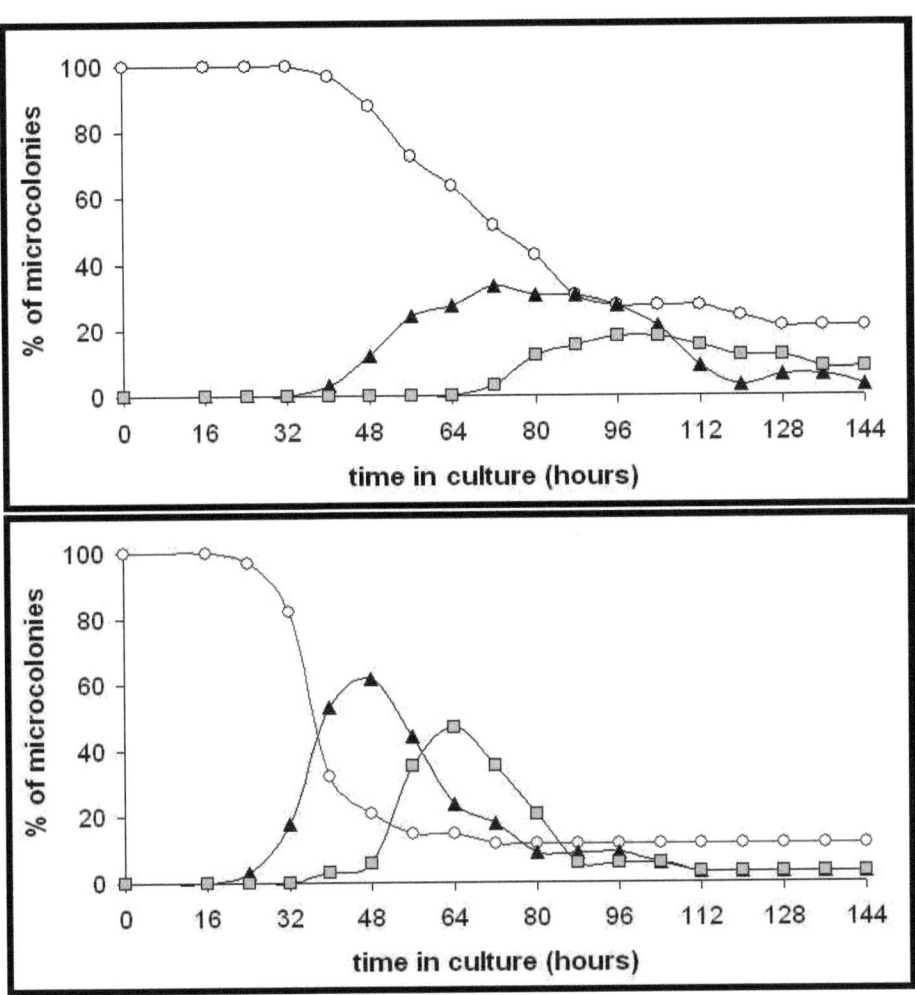

Figure 2.2.: *Microcolony progression analysis of AHAC from one characteristic donor (donor A) expanded in CTR (first picture) or TFP (second picture) medium. Cells were plated at a low density in culture flasks as described in Material and Methods. Every 8 hours, number of cells per microcolony were counted under phase contrast microscopy. The number of a given microcolony type observed is expressed as a percent of the total observed. Microcolonies with three cells were omitted for simplicity.*

area, since the medium supply is assumed abundant and constant. Hence, the value of K corresponds to the maximal number of cells in the monolayer which can be reached at confluence; note that $N(t)$ tends to K as t goes to infinity for any initial value $N_0 > 0$. As the growth can be considered exponential during the first time interval $[0, G_T)$, when $N(t)$ is still very small, the normalization constraint on $\psi(t)$ did not require a further adjustment. Note that equation (2.3) has two steady states, $N = 0$ and $N = 1$. Perturbing it about the state $N = 1$ one finds that this steady state is linearly stable. On the other hand, when we perturb the state $N = 0$, we find the linearized equation $dN/dt = N(t - G_T)$. Upon making the ansatz $N(t) = C \exp(-\rho \lambda G_T)$, we find that the solution to the transcendental equation $\lambda = \exp(-\rho \lambda G_T)$ may have complex solutions in addition to one real positive solution. However, for values of G_T which are about 1, they all turn out to have negative real parts, and therefore the corresponding oscillatory solutions to equation (2.3) are decaying. Hence, we expect the solution to have a non-oscillatory monotone increasing shape from $N = 0$ to the stable state $N = 1$ [37].

Numerical methods

The task of parameter estimation is one of minimizing, in a least-squares sense, an objective function based on a vector of unknown parameters p and sample data t_i, $N_i = N(t_i)$, for $ti = 1, ..., M$. Given an initial value $N(t_0) = N_0$ and an initial function $\psi(t)$ for t in $[-G_T, 0)$, each set of parameter values defines a solution $N(t) = N(t; p)$ for $t \geq 0$, where $p = [\rho, G_T, K]$. We took as N_0 our first experimental data at time $t = 0$. To find the global best-fit parameter values p^* to the data, the initial guess must be su ciently close to p^*. The microcolony tests provided us with a good initial estimate for G_T. To compute $N(t_i, p)$, the DDE is solved with an adaptive fourth-order Runge-Kutta method [24]. The nonlinear optimization problem is solved by the Gauss-Newton method, combined with the Armijo rule for an optimal step length [29].

2.2.3. Statistical analysis

Statistical evaluation was performed using SPSS software version 7.5 software (SPSS, Sigma Stat). Values are presented as mean ± standard deviation (SD). Dierences between cultures in CT R and TFP medium of cells from the same donor were assessed by Student's t-tests for independent samples, after confirming the normality of the populations by skewness and kurtosis. Dierences among donors were assessed by Mann Whitney tests for independent samples. P values less then 0.05 were considered to indicate statistically significant dierences.

2.3. Results

2.3.1. Growth curves

Morphologically, CTR-expanded AHAC were flattened and spread, while TFP-expanded cells were generally smaller with a more elongated, spindle-like shape (Fig 1.1). The growth curves of AHAC from all donors were typically sigmoidal (Fig. 2.2.1): after a lag period of about 3-4 days, cells multiplied exponentially until day 9-10, when they reached the plateau phase. The density of cells counted in the presence of TFP medium was significantly higher than in CTR medium starting from day 5 and was 4.2-fold higher at day 12.

2.3.2. Microcolony tests

For each experiment, 20-60 microcolonies per flask were identified and the number of cells per colony was counted every 8 hours for a total time of 144 hours. The percentages of microcolonies containing

1, 2, or 4 cells were derived at each observation and used to generate microcolony profiles, as shown in Fig. 2.2.1. In CTR medium, the percentage of 1-cell microcolonies declined slowly, reaching a plateau of around 20slow increase in the percentage of 2-cell and 4-cell microcolonies. In the presence of TFP, the percentage of 1-cell microcolonies declined to 1060 hours of culture, due to the rapid appearance of microcolonies with progressively increasing cell numbers (Fig. 2.2.1). Remarkably, the percentage of microcolonies containing more than 4 cells was higher than 50% only at 115 ± 17 hours in CTR medium, but already at 80 ± 8 hours in TFP medium. The collected data were then used to calculate the following kinetic parameters related to AHAC growth (Table 2.1):

- Time of first cell division t_{cd} was highly variable (16-96 hours) even among cells from the same donor and cultured in the same medium, indicating large heterogeneity of dierent AHAC subpopulations. Despite these variations, the mean t_{cd} (T_{cd}) was significantly shorter (1.4-fold, corresponding to 14.8 hours) if cells were cultured in TFP, as compared to CTR medium. In order to further quantify dierences between CTR- and TFP-expanded AHAC, microcolonies were arbitrarily divided into the following three groups: group I for $t_{cd} \leq 16$ hours, group II for t_{cd} between 17 and 32 hours, and group III for $t_{cd} \geq 33$ hours. As compared to CTR-expanded AHAC, those expanded in TFP contained a statistically significant higher fraction of microcolonies in group I (0.7% vs 9.1%) (Fig. 2.3.2, A).

- Percentage of quiescent cells Q_c was 1.7-fold lower in AHAC cultivated in the presence of TFP as compared to CTR medium, indicating that the growth factor mix induced a significant increase in the proportion of mitotically active cells.

- Generation time g_T was highly variable (12-72 hours) even among microcolonies from the same donor and cultured in the same medium, again underlining a large heterogeneity of dierent AHAC subpopulations. Despite these variations, the mean g_T (G_T) was significantly shorter (1.2-fold, corresponding to 4.8 hours) in AHAC cultivated in the presence of TFP as compared to CTR medium. Microcolonies were again classified into the following three groups: group I for $g_T \leq 16$ hours, group II for g_T between 17 and 32 hours, and group III for $g_T \geq 33$ hours. As compared to CTR-expanded AHAC, those expanded in TFP contained a statistically significant higher fraction of microcolonies in group I (11.9% vs 25.4%) and a lower fraction of microcolonies in group III (23.4% vs 11.7%) (Fig. 2.3.2, B).

- Fraction of cells that divide per g_T $f_c(T)$ is an index of the exponentiality of the cell growth, calculated for each microcolony ($f_c(T)$ equals 1 for an ideal exponential growth, and 0.5 for a linear population growth). The mean $f_c(T)$, $F_c(T)$, was similar in CTR- and TFP-expanded populations and close to 1, indicating a common exponential pattern of cell growth.

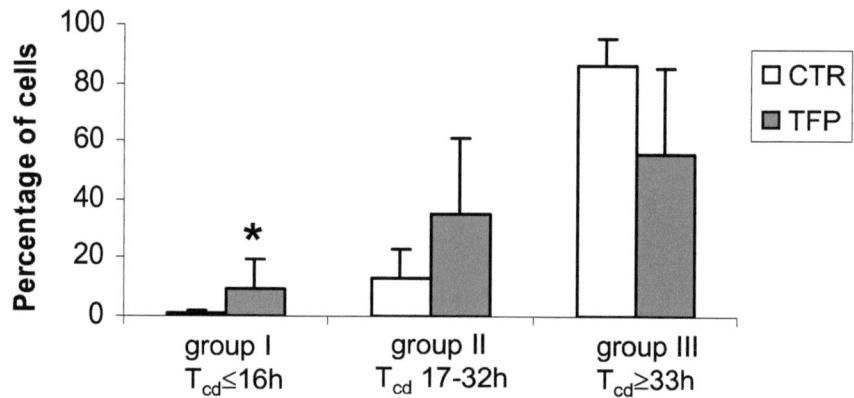

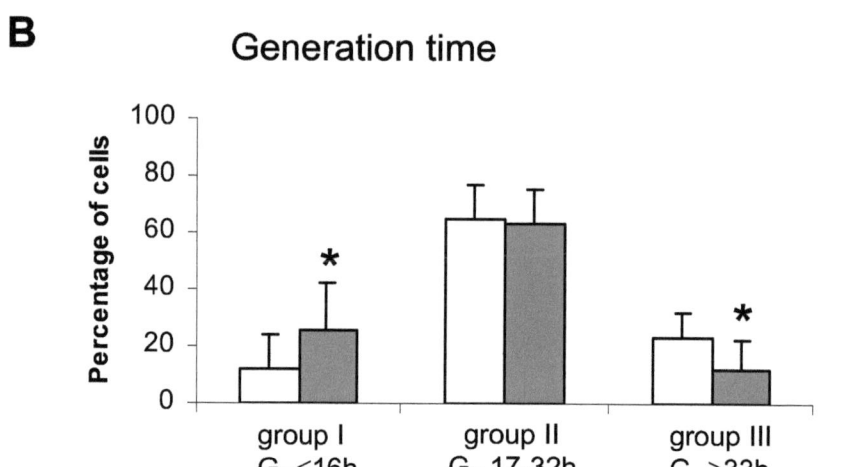

Figure 2.3.: *Time of first cell division (A) and generation time (B) of cells expanded in CTR or TFP medium, following subdivision in arbitrary groups.*

Table 2.1.: Microcolony growth parameters

	Exp cond	Donor A mean ± SD[a]	Donor B mean ± SD[a]	Donor C mean ± SD[a]	Average mean ± SD[b]
t_{cd} (hours)	CTR	48.6 ± 19.7	64.3 ± 22.6	53.6 ± 21.9	55.5 ± 8.0
	TFP	32.6 ±15.5*	36.3 ±11.3*	53.3 ± 16.9	40.7 ±11.0*
Q_c	CTR	19.0	21.2	0.30	0.26
	TFP	7.7	11.8	0.34	0.30
g_t (hours)	CTR	29.2 ± 6.4	25.5 ± 10.4	31.1 ± 13.0	28.6 ± 2.9
	TFP	25.1 ± 11.1	20.6 ± 4.9*	25.6 ± 12.2*	23.8 ±2.7*
F_c	CTR	0.86 ± 0.08	0.87 ± 0.14	0.85 ± 0.14	0.86 ± 0.01
	TFP	0.85 ± 0.11	0.90 ± 0.15	0.87 ± 0.15	0.87 ± 0.02

([a])*Summary of growth parameters estimated by microcolony tests.*
Parameters are reported as mean ± SD of microcolonies within the same donor.
([b])*Parameters are reported as mean values calculated for each donor ± SD.*
(*)$P < 0.05$ *from CTR medium.*

2.3.3. Numerical simulations

To determine G_T and ρ, the exponential delay model 2.2 was first fitted to the measurements from the three experiments during the exponential phase only. The resulting values for G_T were similar to those obtained from the microcolony test (Table 2.1 and 2.2). Between the two expansion conditions (in CTR or TFP medium), only a slight dierence in G_T was observed (average variation = 9.4%) , while in general there was a large dierence in ρ (average variation = 34.3%) (Table 2.2). This can be explained by the fact that ρ represents the overall proliferation rate taking into account the number of quiescent cells and the time of first cell division, parameters that have been shown to have higher values in CTR medium.

Table 2.2.: Growth parameters calculated by (2.3)

	Donor A			Donor B		Donor C	
	G_T (hours)		D_T (hours)	G_T (hours)	D_T (hours)	G_T (hours)	D_T (hours)
CTR	27.0	0.86	33.7	19.5	0.75	33.3	29.8	1.10	30.1
TFP	24.0	1.60	22.1	17.0	1.10	24.5	28.0	1.40	25.4

	Average	G_T (hours) mean ±SD	mean ± SD	D_T (hours) mean ± SD
	CTR	25.4 ± 5.3	0.90 ± 0.18	32.5 ± 2.0
	TFP	23.0 ± 5.6	1.37 ± 0.25	24.0 ± 1.7

In the study of cell growth dynamics, another typical parameter is the doubling time D_T. For a simple exponential model (without delay), D_T is constant and equal to $\ln(2)/\rho$. However, in our exponential model with delay, the doubling time varies with time. Nevertheless, it reaches an asymptotic limit (D_T)

Table 2.3.: Carrying capacity K calculated by (2.3)

	Donor A	Donor B	Donor C	Average mean ± SD
CTR	10.1	12.6	8.7	10.5 ± 2.0
TFP	40.4	37.0	38.0	38.5 ± 1.8

Values of K are fitted by the delay model (2.3) to the experimental data. In the last columns, mean values calculated from each donor ± SD are reported.

at later times, as transients during the initial stage are dissipated. Clearly the asymptotic value D_T depends on G_T and ρ. As the behavior of the delay model tends to that of an exponential model at later times, we can seek for a solution of 2.2 of the form

$$N(t) = C \exp(\rho \lambda t), \qquad (2.4)$$

where C and λ are some positive constants. We introduce 2.4 into 2.2 which leads to the transcendental equation

$$\lambda = \exp(-\rho \lambda G_T). \qquad (2.5)$$

By solving this equation for λ with Newton's method for different values of G_T and ρ, we can calculate $D_T = \ln(2)/(\rho \lambda)$. In Fig. 2.3.4 the relationship between D_T and G_T for a typical value of ρ is shown. We observed that D_T increased with increasing G_T in a nonlinear way. In Table 2.2 the values of D_T, extrapolated for each donor and expansion condition, are reported. Interestingly, while G_T varied substantially between different donors, D_T remained almost similar in all experiments (% variation: about 20% for G_T and 7% for D_T). Since the fit to the exponential delay model yielded a good estimate of G_T, as confirmed by the microcolony tests, these values were then used in the logistic delay model to obtain the carrying capacity, K. Consistently with the experimental data, K was four times larger in the presence of TFP than in CTR medium, probably due to the efficiency of space occupation (Table 2.3). A reduction of the error between a standard logistic model and our delay logistic model 2.3 was observed (the mean error in the former was 1.17 and 1.06 respectively for CTR and TFP medium, but only 0.69 and 0.96, respectively, in the second); hence, the new model approximates better the observations. In Fig. 2.3.4 we show the solutions obtained by fitting the delay logistic model 2.3 to the experimental data for each donor in the two expansion conditions.

2.3.4. Difference in the growth kinetic between AHAC at different passages in culture

An intriguing question is whether prolonged expansion in the presence or absence of TFP could induce an enrichment of AHAC populations with the fastest growth capacity. To address this question, AHAC from one donor (donor C) were expanded in CTR or TFP medium either for one (P1 cells) or two (P2 cells) passages, corresponding respectively to 1.4 or 13.3 doublings for CTR and 2.5 or 17.2 doublings for TFP, and then assessed using the microcolony test in combination with the developed mathematical model. Unexpectedly, no difference was observed in the G_T measured using P1 or P2 cells expanded in CTR or TFP medium (Table 2.4), and the percentage of fast subpopulations (group I) in P2 cells was lower than in P1 cells (2.2 and 4.4-fold respectively for CTR and TFP) (Fig. 2.3.5). The accuracy of the experimentally determined G_T was confirmed by the fact that the mathematical model was able to predict the effective temporal growth in cell number only if the measured G_T, but not a shorter G_T, was given as input (Fig. 2.3.5). Interestingly, as compared to P1 cells, P2 cells had a shorter T_{cd} (1.3-fold, corresponding to 11.7 hours, in CTR medium and 2.1-fold, corresponding to 27.8 hours, in TFP medium)

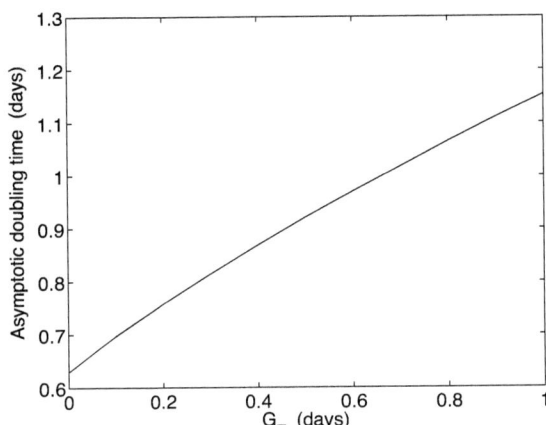

Figure 2.4.: *Relationship between generation time G_T and doubling time D_T. Values of these parameters plotted in this diagram were obtained from cells derived from donor B and expanded in TFP (= 1.1).*

and a lower Q_c (6.9-fold and 8.2-fold respectively for CTR and TFP medium) (Table 2.4). Moreover, prolonged expansion induced an enrichment of cells with short T_{cd} (groups I and II) especially using TFP medium (Fig. 2.3.5).

Table 2.4.: Microcolony growth parameters: dierences between P1 and P2

	CTR		TFP	
	P1	P2	P1	P2
t_{cd} (hours)	53.6 ± 21.9	41.9 ± 18.3*	53.3 ± 16.9	25.5 ± 14.7*
Q_c	30.3	4.4	21.2	2.6
g_T (hours)	31.1 ± 13.0	27.5 ± 9.7	25.6 ± 12.2	26.1 ± 7.9
f_c	0.85 ± 0.14	0.88 ± 0.13	0.87 ± 0.1	0.85 ± 0.12

Summary of growth kinetic parameters derived for cells from patient C cultured for one passage (P1) or two passages (P2) in CTR or TFP medium.
(*)$P < 0.05$ *from P1 cells.*

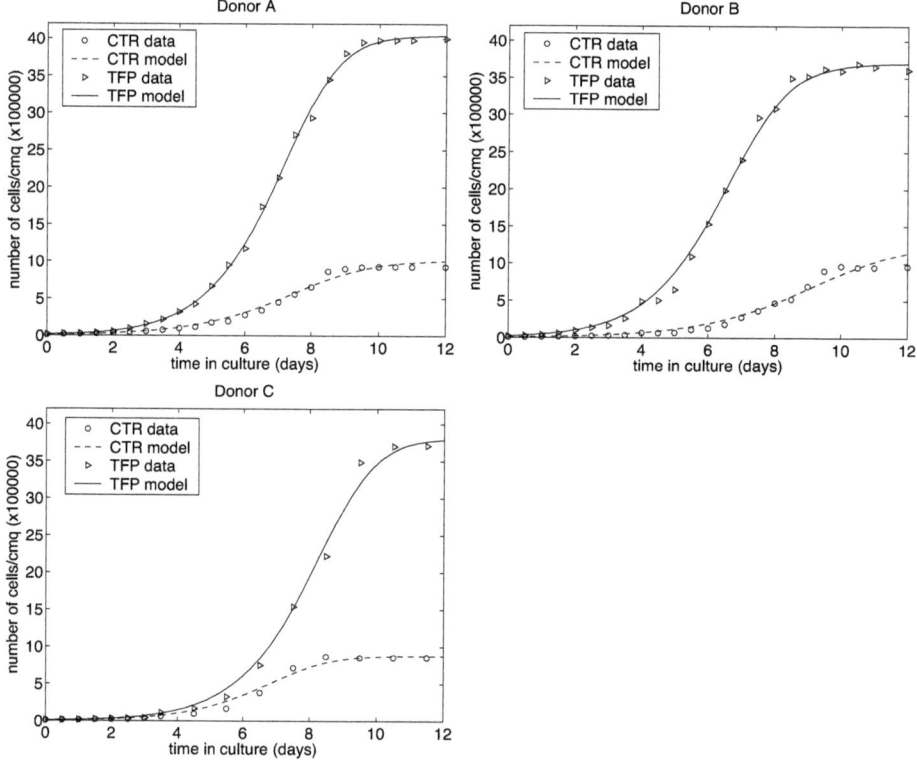

Figure 2.5.: *Experimental and predicted growth curves of cells from donors A,B and C.*

2.3.5. Discussion

In the present study, we used a combination of microcolony tests and a newly developed mathematical model, combining logistic growth with time delay, to (i) measure the kinetic parameters of AHAC, (ii) capture the entire growth process, and (iii) investigate the specific effects of the growth factor combination TFP on cell proliferation. We found that TFP medium increases the number of chondrocytes in monolayer culture by reducing (i) the percentage of quiescent cells (Q_c), (ii) the mean time required for single cells to enter the first division (T_{cd}), and (iii) the mean cell division time of cells (generation time, G_T). Our mathematical model confirmed the value of G_T and provided estimates for the carrying capacity of the system (K) and the proliferation rate (ρ), which were respectively 3.7- and 1.5-fold higher in chondrocytes cultured in TFP. Moreover, our results for chondrocytes from the same donor at different passages in culture indicate that prolonged expansion does not increase the fraction of the fastest proliferating AHAC, but rather the fraction of AHAC with a higher propensity to initiate duplication, particularly in the presence of TFP. In ([41]) was first used the microcolony test as a tool to investigate changes in growth properties of mouse mammary epithelial cell lines under condition inducing elevated p53 expression. We also adopted the microcolony test to study the growth kinetic of AHAC and its modulation due to the presence of TFP medium. As compared to CTR-expanded AHAC, those expanded in TFP contained a higher fraction of cells with short generation time (less than 16 hours) and a lower fraction of AHAC with high generation time (more than 32 hours). However, based on our experimental data, we could not determine whether all the cells in culture or only specific cell subpopulations responded to TFP by reducing their g_T. Interestingly, the overall growth pattern of chondrocytes was not altered by TFP, as indicated by the fact that identical values of the index of the exponential nature ($F_c(T)$) were obtained in chondrocytes cultured with and without growth factors. The finding that $F_c(T)$ was close to 1 in both conditions allowed us to use an exponential model for the initial phase of cell growth. One striking observation was that the time of first division and the generation time greatly differed among microcolonies derived from different cells of the same patient, in agreement with a previously described large heterogeneity of different AHAC clonal subpopulations ([7]). To capture the non-instantaneous and asynchronous first cell division, a delay model had to be adopted. Using a delay model it is in fact possible to distinguish between G_T, a characteristic of a single cell, and the doubling time (D_T), a global feature of the whole cell population. G_T is clearly shorter than the doubling time, since it does not take into account the quiescent cells and the delay in the first cell division. We observed that D_T always tends to an asymptotic value, different for different values of G_T, and more interestingly, that there is a relationships between G_T and the asymptotic value of D_T; this relationship depends on the proliferation rate (ρ), but not on the initial cell number. Once having estimated G_T and ρ by fitting our model to the data, we can thus extrapolate the correspondent value of D_T from the graph. Our model characterized cell growth during the entire experiment, until confluence. This allowed us to estimate the number of cells at confluence (K), which was remarkably different between the two expansion conditions (CTR or TFP medium). This difference can be explained by the smaller and more elongated cell shape induced by TFP. Further studies have to be performed to assess whether cells cultured in TFP medium have also an increased tendency to migrate, which would lead to a more efficient occupation of the available space. In this study we also aimed at determining whether prolonged expansion in the presence or absence of TFP induces an enrichment of the populations with the highest growth capacity. Unexpectedly, we observed that P1 and P2 chondrocytes divided with unchanged mean G_T, and, more interestingly, that the fraction of fast cells (G_T lower than 16 h) decreased dramatically from P1 to P2 chondrocytes. These differences were more pronounced if cells were expanded in TFP medium. Since replicative aging occurs during *in vitro* cell expansion and the senescence-mediated phenomena become more evident in cells undergoing elevated population doublings ([26]), it is possible to speculate that cell senescence following prolonged expansion masked a possible selection of the fast chondrocytes. On the other hand, P2 chondrocytes had shorter T_{cd}, larger fraction of cells with short t_{cd}, and lower Q_c than P1 chondrocytes especially if cells

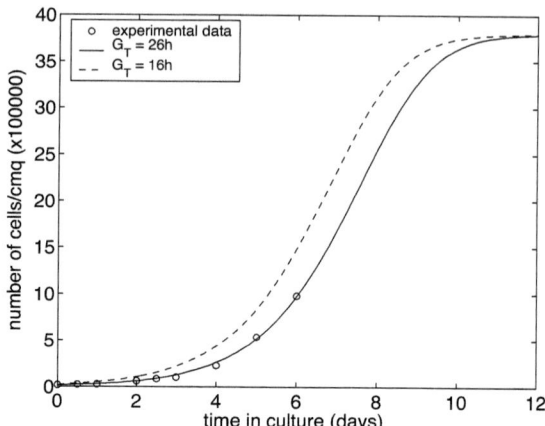

Figure 2.6.: *Growth curves of P2 cells from donor C obtained applying G_T of 16 or 26 h in the logistic model. The circles indicate the experimentally determined number of cells. Applying a G_T of 16 h in the model clearly resulted in an inaccurate fit.*

were expanded in TFP medium. This result indicates that prolonged expansion, particularly in medium containing TFP, might induce a selection of chondrocytes with a higher propensity to initiate duplication. In the present work, we have studied the influence of the growth factor combination TGFβ1, FGF-2, and PDGF-BB on the growth kinetic of adult human articular chondrocytes using both microcolony tests and a mathematical model. The described approach could be adopted to quantitatively assess the growth of other cell types, cultured under dierent experimental con ditions.

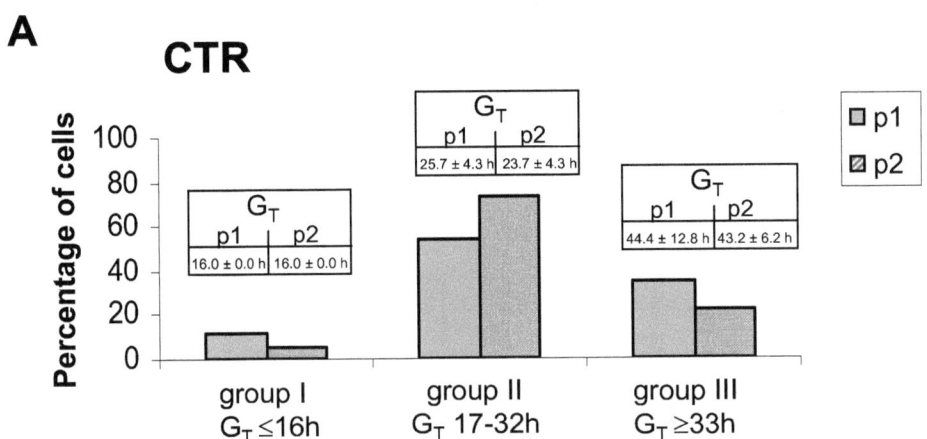

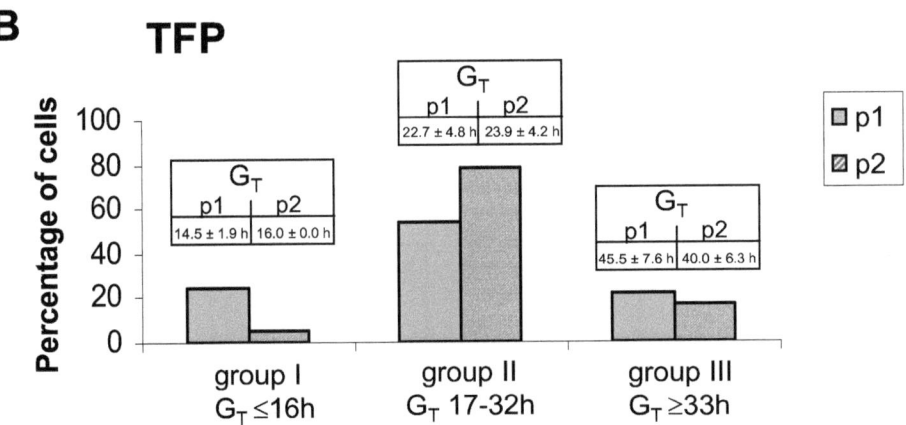

Figure 2.7.: *Generation time G_T in P1 and P2 cells expanded in CTR (A) or TFP (B) medium, following subdivision in arbitrary groups.* $^*P < 0.05$ *from P2 cells.*

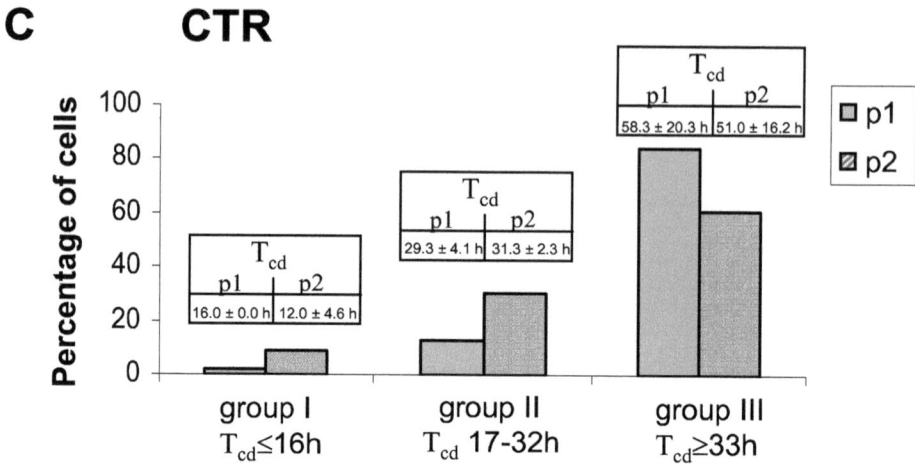

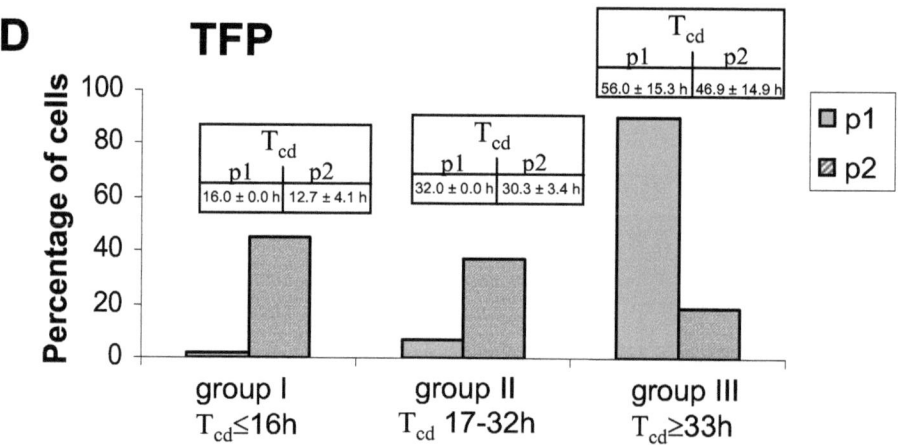

Figure 2.8.: *Time of first cell division Tcd in P1 and P2 cells expanded in CTR (C) or TFP (D) medium, following subdivision in arbitrary groups.* $^*P < 0.05$ *from P2 cells.*

3. Dynamic Formation of Oriented Patches in Chondrocyte Cell Cultures

The content of this chapter has been accepted for publication on the Journal of Mathematical Biology. In chapter 4 the model presented here is analyzed more in details.

Growth factors have a significant impact not only on the growth dynamics but also on the phenotype of chondrocytes (Barbero et al. , J. Cell. Phys. 204, pp. 830-838, 2005). In particular, as chondrocytes approach confluence, the cells tend to align and form coherent patches. Starting from a mathematical model for fibroblast populations at equilibrium (Mogilner et al., Physica D 89, pp. 346-367, 1996), a dynamic continuum model with logistic growth is developed. Both linear stability analysis and numerical solutions of the time-dependent nonlinear integro-partial dierential equation are used to identify the key parameters that lead to pattern formation in the model. The numerical results are compared quantitatively to experimental data by extracting statistical information on orientation, density and patch size through Gabor filters.

3.1. Introduction

In recent years, therapies for damaged tissue have experienced great progress through the possibilities oered by new methods of tissue engineering [31]. In particu lar, this emerging field holds great promise for the regeneration of tissues with limited intrinsic repair capacity like adult articular cartilage. Most procedures pursued in tissue engineering start with a biopsy containing few cells, which are expanded *ex vivo* making use of cytokines. Those are then injected into a patient to grow tissue *in vivo*. Alternatively, scaolds are used to provide mechanical support and structu re for the tissue to be grown *in vitro* before its implantation on the defect. Those procedures may involve tissue replacement using donor tissue or autologous cells for *in vitro* cell-culture expansion, in order to regenerate tissue that matches the patient's native tissue.

Much research has already gone into the impact of combinations of growth factors on the proliferative capacity for a range of cell types, such as pancreatic endorcine cells, neural progenitor cells, muscle-derived stem cells [8, 9, 14, 45] and chondrocytes [5–7, 28]. For muscle-derived stem cells and chondrocyte cell cultures, mathematical models that establish characteristic kinetic parameters, such as the fraction of dividing cells and mean cell division-time have been developed [6, 14]. In addition, a logistic delay-dierential model for proliferating chondrocyte cells was used to further include the eect of contact inhibition of proliferating cells upon confluence [6].

Apart from the impact of growth factors on kinetic parameters, an important focus of research concerns the eect of cytokines on the phenotype of individual cells a nd the resulting organizational structure of the cell culture. Both will influence the mechanical properties of the engineered tissue, which in the case of cartilage, is intended to sustain tensile stresses and compressive loads, just as native tissue does [40]. Therefore, it is important to understand the underlying processes leading to the formation of large-scale patterns of an evolving cell culture. Identifying the relevant parameters that control these structures is the focus of the present study, combining experimental and mathematical methods [10].

In [6], adult human articular chondrocytes (AHAC) were isolated from cartilage biopsies and then cultured in the presence of a combination of growth factors. The individual cells assume a phenotype that closely resembles fibroblasts and eventually self-organize into regions of aligned cells, making up the monolayer of the cell culture at confluence [7]. This phenomenon has been observed before for various cell types. In principle there are a number of mechanisms that may control the formation of such patterns, ranging from chemical, adhesive or other mechanical gradients, see e.g. Trinkaus [48] for an early but instructive discussion in the context of morphogenesis.

Even in the absence of exterior influences, however, Elsdale [18] discovered that proliferating fetal lung fibroblasts form parallel arrays during *in vitro* cell expansion. Similar results were found for BHK fibroblasts in the experimental study by Erickson [19]. In [18] Elsdale argued that the intrinsic property of fibroblasts is to move, unless prevented to do so by the environment, and hence that patterns form solely due to direct cell-cell interactions to enable maximal motility. Under the assumption of contact inhibition, Erickson[19] concluded from a series of cell-cell contact experiments that if the lamellipodium of a cell in ru ing mode contacts another cell at a certain angle, the d irection of motion changes depending on that part of the leading edge of the lamellipodium which made contact and where ru ing is stopped. This mechanism is employed by Erickson to explain the existence of a critical angle above which cells cease to align. This critical angle seems to dier for diere nt cell types, e.g. about 20° for fetal lung fibroblasts and approximately 50° for BHK cells. For fibroblasts the leading edge of the lamellipodium is much narrower than for the BHK fibroblasts. Hence, except for rather narrow contact events, motion will halt (else cells may even criss-cross other cells). Moreover, similar behaviour is observed for contact events of already established arrays of aligned cells. This behaviour is eventually reflected in the resulting patterns at confluence.

Mathematical modeling of the dynamical process of array formation of aligned cells started with the work by Edelstein-Keshet and Ermentrout [16]. The continuum models derived for pattern forming cell cultures assume random spatial and orientational distributions of the cells that are attracted (repulsed) and change their direction of motion in response to cell-cell interactions. Here the cell density depends on time, two-dimensional physical space and the angle of orientation. The range of interaction is kept small in order to model the local character of cellular interactions. Apart from terms modeling the random motion in physical and angular space, the model includes a term that describes the probability of alignment of cells as a response to cell-cell contact, which vanishes outside the range of angles known to lead to alignment. In subsequent articles the resulting system of integro-dierential equations for free cells and cells already bound to an array are discussed in various limiting cases and analysed with respect to their stability about the homogeneous state [34]. Similar models were also used for other pattern forming processes such as swarming or the dynamics of actin binding fibers [11, 34, 36].

Here we extend these models by including time-dependent logistic growth to account for the later stages of *in vitro* chondrocyte cell expansions. In fact, one important aspect of our study is to enable a direct comparison with our experimental results in section 3.2.1. The analysis of the experimental results and, in particular, the classification of the cells within angular space is realized by using two-dimensional Gabor filters [13] for the experimental images and is described in section 3.2.3. In section 3.3, we present our mathematical model, which consists of a time-dependent nonlinear integro-partial dierential equation. We use standard finite dierences for the numerical discreti zation in space, for the time discretization we use explicit Chebyshev methods that circumvent the crippling stability restrictions of standard explicit Runge-Kutta methods – see the Appendix. In section 3.4, we investigate the stability of the solution via a linear stability analysis about the homogeneous state and compare those findings to the results of the full nonlinear model. Finally, quantitative comparisons with experimental data are performed in section 3.5.

3.2. Biological background

In appendix C the method used and the experiments are explained in details.

3.2.1. The impact of growth factors

Depending on the cell type and the specific growth factors used, cytokine-induced proliferation of cells can generally be characterized by one or more parameters, such as a shorter cell division time, a shorter time until first cell division, or lower percentage of remaining quiescent cells [6, 14, 15]. Those key parameters can be obtained, for instance, by combining a logistic delay-dierential model with the results from specific micro-colony experiments [6]. From that model, Barbero et al. established in the case of adult human articular chondrocytes (AHAC) expansion in a medium supplemented with the growth factor combination TGFβ-1, FGF-2 and PDGF-BB (TFP) that the time of first cell division is about 1.4 times shorter and the percentage of quiescent cells about 1.7 times smaller than in the absence of TFP.

Further characteristics observed in experiments [6] concern the elongated shape the cells assume when cultured in a medium with TFP. During the sigmoidal growth of the cell culture, individual cells are initially oriented at random. As the population approaches confluence, cells tend to locally align and form coherent structures. Those spatial patterns appear highly irregular while individual patches greatly vary both in shape and size, without clear boundaries between them – see Fig. 3.1.

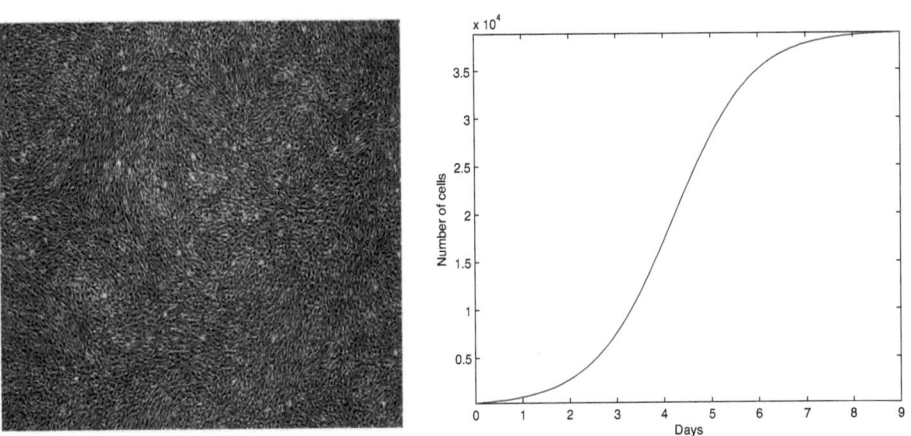

Figure 3.1.: *AHAC cultured with TFP at confluence, day 9 (left). Sigmoidal evolution of the number of cells vs. time (right)*

3.2.2. Cell culture: isolation and expansion

To monitor patch formation and obtain quantitative experimental data on diusion constants, we track the motion of an ensemble of individual AHAC cells up to confluence. Here full-thickness human articular cartilage samples were collected from the femoral lateral condoyle of two individuals (patient A: male, 18 years old, patient B: male, 66 years old), with no history and no radiographic signs of joint disease, after informed consent and in accordance with the local Ethical Commission. Adult human articular chondrocytes (AHAC) were isolated using 0.15% type II collagenase for 22 hours and cultured for one passage in Dulbeccos modified Eagles medium (DMEM) containing 10% foetal bovine serum, 4.5 mg/ml D-Glucose, 0.1 mM nonessential amino acids, 1 mM sodium pyruvate, 100 mM HEPES buer, 100 U/ml penicillin, 100 ï£·g/ml streptomycin, and 0.29 mg/ml L-glutamine and supplemented with the 1 ng/ml of Transforming Growth Factor-b1 (TGF-b1), 5 ng/ml of Fibroblast Growth Factor-2 (FGF-2) and 10 ng/ml of Platelet-Derived Growth Factor-BB (PDGF-BB) (growth factor medium, TFP) in a humidified $37^oC/5\%$ CO2 incubator as previously described [6]. When cells were approximately 80% confluent, first passage (P1) cells were rinsed with phosphate buered salin e, detached using 0.05% trypsin/0.53mM EDTA and frozen in complete medium containing 10% dimethylsulfoxid. AHAC after thawing were then used for the studies described below.

Monitoring of cell expansion until confluence
AHAC were re-plated in two wells of a 6 well plate at a density of 10000 cells/cm^2 and cultured in growth factor medium up to 10 days in a humidified $37^oC/5\%$ CO2 incubator with daily culture medium change. AHAC cultures were monitored by phase contrast microscopy and pictures were taken from random areas of the wells each day.

Study of cell movement
AHAC were re-plated in a 6 well plate at dierent densities, w hich were 200, 3000, 10000, 15000 and 20000 cells/cm^2, and cultured in growth factor medium for 1 day in a humidified $37^oC/5\%$ CO2 incubator. Next, the plate was transferred to the incubator of the Olympus system. From a time-lapse microscope we obtained a sequence of frames which we used to monitor cell motion. Snapshots were taken at 15 minute intervals, which corresponds to an average travel distance of $9\mu m$, to provide su cient space-time resolution compared to the typical length of a cell ($50\mu m$). With the software analySIS^D we performed a manual tracking in all five wells (Fig. 3.2) to follow about 100 cells for each density, for 12 hours, a duration that allowed us to neglect cell doubling.

Spatial diusion
To estimate the spatial diusion, we performed experiments at various densities (i.e. 200, 3000, 10000, 15000, 20000 cells/scm) and for each density we manually tracked individual cells in three dierent areas of the well. Assuming Brownian motion, the diusion co e cient D is related to the mean square displacement, $<X^2>$, through the relation $<X^2> = 2Dt$. A linear least-squares fit of the time evolution of the mean square distance then yields D. From those estimates at varying density, shown in table C.1, we obtained the constant average diusion coe ci ent $D = 0.29\,\mu m^2/s$.

cells/scm	200	3000	10000	15000	20000
coe D1	0.31	0.30	0.37	0.31	0.31
coe D2	0.18	0.23	0.32	0.30	0.26
coe D3	0.23		0.40	0.34	0.30
mean ± SD	0.24 ± 0.07	0.26 ± 0.04	0.36 ± 0.04	0.31 ± 0.02	0.30 ± 0.03

Table 3.1.: Estimates of the diusion constant D at three dierent locations inside the well and at varying density, together with the mean values ± SD.

Figure 3.2.: *Tracking of individual cells at density $20000/cm^2$.*

3.2.3. Image analysis of alignment

Standard image segmentation algorithms proved unable to distinguish between individual cells and the background. Thus to identify patches of alignment and estimate their size quantitatively, both in the numerical simulations and in the experiment, we proceed in two steps. First, we apply a special class of filters to images such as Fig. 3.1 that reveal the dominant local axis of orientation. Second, we estimate the average size of cell clusters with a common orientation through a discrete statistical measure, which is then used to compare numerical simulations with experiments.

To classify cells according to their orientation, we opt for Gabor filters [21, 23] which consist of a local Gaussian kernel of width σ, multiplied by a plane wave with distinct orientation θ and frequency ω:

$$G(x', y') = \exp\left\{-\frac{1}{2}\left(\left(\frac{x'}{\sigma}\right)^2 + \left(\frac{y'}{\sigma}\right)^2\right)\right\} \cos(2\pi\omega x')$$

$$x' = x\cos(\theta) + y\sin(\theta), \quad y' = y\cos(\theta) - x\sin(\theta),$$

where unit length in x (or y) corresponds to a single pixel. The typical width $\sigma = 12$ and frequency $\omega = 0.08$ for an array of aligned cells were determined a priori and remained fixed in all further analysis – see Fig. 3.3. Hence Gabor filters locally respond to patterns with spatial frequency ω and orientation θ, within a subregion of size σ. Their two-dimensional extension is commonly used in image analysis and

 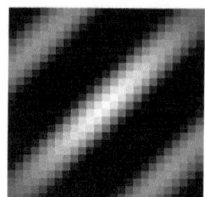 

Figure 3.3.: *Two-dimensional Gabor filters with frequency* = 0.08, *scale* = 12, *and orientations* = 0, / 4, / 2.

computer vision; they were also proposed as a model for the spatial summation properties of simple cells in the visual cortex [13].

To any image we apply a suite of Gabor filters for varying orientation at 45° intervals and assign to each pixel location (i,j) a distinct color c_{ij} that corresponds to the highest filter response. Hence c_{ij} reflects the dominant orientation at location (i,j), and cells aligned with that particular orientation are thus revealed, as shown in Fig. 3.4.

Next, we estimate the typical cluster in a filtered image, such as in Fig. 3.4, either from experiment or numerical simulation. To do so, we assign to each pixel (i,j) the value $p_{ij}(s) = 1$ if it belongs to a cluster of size s, that is if at least 50% of the points within distance s are of the same color; else, we set $p_{ij}(s)$ to zero. Summation over all pixels yields an estimate $r(s)$ of the number of pixels belonging to a cluster of size s as

$$r(s) = \sum_{ij} p_{ij}(s), \quad p_{ij}(s) = \begin{cases} 1 \\ 0 \end{cases}. \tag{3.1}$$

The intersection of $r(s)$ with the s-axis yields a reliable estimate for the typical patch size, i.e. the largest cluster size, as illustrated with synthetic black and white data in Fig. 3.5. Moreover, comparison of the left and right frames in Fig. 3.5 demonstrates that the intersection of $r(s)$ with the s-axis is rather insensitive to added random noise.

3.3. Mathematical Model

In appendix B.1 there is a detailed explanation of how this equation can be derived.

3.3.1. Formulation

Starting from the pioneering works of Edelstein-Keshet et al. [16, 34, 35], we now build a continuum model to describe the time evolution of a cell population of density $C(\theta, \vec{x}, t)$ in angle θ and two-dimensional space $\vec{x} = (x, y)$ at time t. During initial times the cells are essentially free to move in space and also turn their axis of orientation at random, similarly to fibroblasts. As the population density increases, however, cells come into contact. In [16, 18, 19] the underlying mechanism responsible for the directional

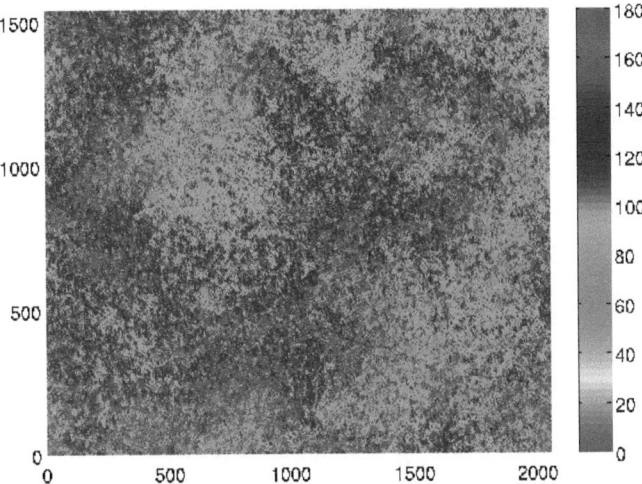

Figure 3.4.: *The effect of Gabor filtering when applied to Fig. 3.1. The color indicates the dominant local direction of alignment.*

motion and the resulting pattern formation is explained solely on the basis of single cell contact events for the case of related fibroblast cell cultures. This mechanism is a form of contact inhibition that cells experience when their lamellipodium touch. Indeed whenever mutual contact occurs within a small angle and hence only a portion of the lamellipodium touches, the cells alter their orientation accordingly and align, as observed by Elsdale [18].

Following [35] we now let $W(\vec{x} - \vec{x}', \theta - \theta')$ denote the rate at which a cell at $\vec{x}'$ and θ' moves to $\vec{x}$ and rotates to θ due to the impact of any surrounding cells. The angular velocity associated with this motion is then given by the gradient of W at angle θ and position $\vec{x}$, due to the cumulative interaction with all other cells:

$$\frac{\partial}{\partial \theta}(W \star C)(\vec{x}, \theta, t) := \frac{\partial}{\partial \theta} \int W(\vec{x} - \vec{x}', \theta - \theta') C(\theta', \vec{x}', t) \, d\theta' d\vec{x}'. \tag{3.2}$$

The gradient of the associated flux $C \, \partial \, (W \star C)$ then induces convective motion towards locations of higher concentration which corresponds to aggregation in space and alignment in angle; both compete with the inherent tendency of cells for random motion modelled by diffusive terms.

Next, the probabilities to align or to aggregate are assumed independent of each other, that is

$$W(\vec{x} - \vec{x}', \theta - \theta') = W_1(\theta - \theta') W_2(\vec{x} - \vec{x}'). \tag{3.3}$$

Moreover, experiments suggest that the probability of alignment W_1 decreases as the relative angle between neighboring cells increases [18], whereas beyond a critical angle α cells no longer align; hence, W_1 must be positive and non-increasing for $0 \leq \theta \leq \alpha$ but become negligible for $\alpha < \theta \leq \pi$. Since

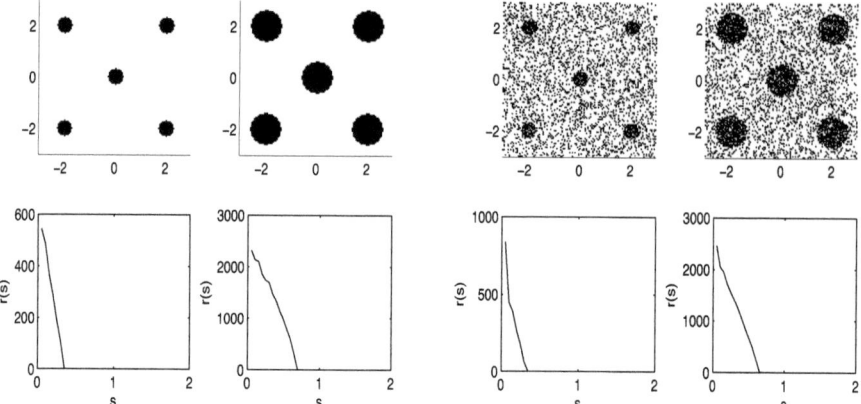

Figure 3.5.: *Estimate of cluster size from two-dimensional synthetic texture images (top left), with corresponding cluster size function $r(s)$ below, as defined in (3.1) (bottom left). The intersection of $r(s)$ with the s-axis yields a robust estimate of the cluster size, even in the presence of added random noise (right).*

clockwise and anticlockwise turns are equally probable, W_1 must also be even. For simplicity, we assume that W_1 is Gaussian with mean zero and standard deviation $\alpha/2$; other choices are possible and discussed in [16]. After normalization, we thus obtain

$$W_1(\theta) = \frac{1}{\alpha\sqrt{2\pi}} e^{-\frac{\theta^2}{2}}. \tag{3.4}$$

Since the strength of cell-to-cell interactions decreases with growing distance [16], we again choose a Gaussian kernel for W_2,

$$W_2(\vec{x}) = \frac{1}{2\sigma^2\pi} e^{-\frac{|\vec{x}|^2}{2}}, \quad \vec{x} \in [-L_x, L_x] \times [-L_y, L_y], \tag{3.5}$$

where L_x and L_y denotes the size of the domain.

Our previous experiments indicate that the growth rate slows down, as the cell density increases locally in space, and that it eventually vanishes when the carrying capacity is reached because of limited space [6]. Therefore we model cell growth by a logistic term with growth rate ρ, where the growth rate reduction is determined by the population density at x and t, that is by the marginal probability density $\int_- C(t,x,y,\theta) d\theta$. The full logistic model, including the divergence of the drift of the cell population

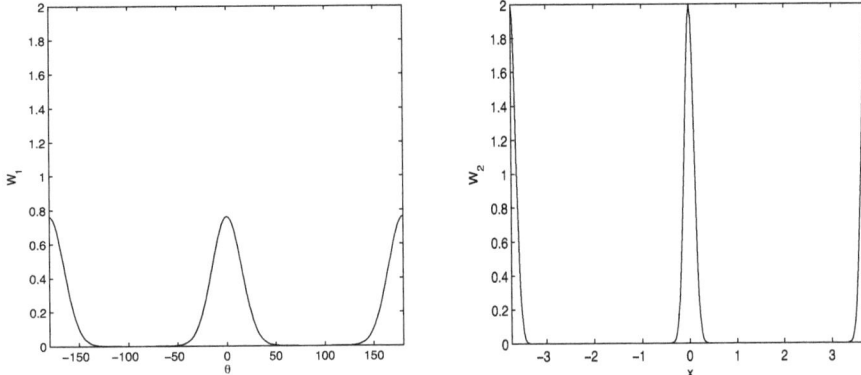

Figure 3.6.: *The angular and spatial cellular interaction kernels from (3.4), (3.5): $W_1(\theta)$ with critical angle $\theta = 50°$ (left), and $W_2(x)$ (right).*

and random motion, previously derived in [35], can then be written as

$$\begin{aligned}
\frac{\partial C}{\partial t} =\ & \epsilon_1 \frac{\partial^2 C}{\partial \theta^2} + \epsilon_2 \left(\frac{\partial^2 C}{\partial x^2} + \frac{\partial^2 C}{\partial y^2} \right) \\
& - \gamma \frac{\partial}{\partial \theta} \left(C \left[\frac{\partial W}{\partial \theta} * C \right] \right) - \gamma \left\{ \frac{\partial}{\partial x} \left(C \left[\frac{\partial W}{\partial x} * C \right] \right) + \frac{\partial}{\partial y} \left(C \left[\frac{\partial W}{\partial y} * C \right] \right) \right\} \\
& + \rho C \left(1 - \frac{L_x L_y}{K} \int_{-}^{} C(t, x, \theta) d\theta \right).
\end{aligned} \tag{3.6}$$

where ϵ_1, ϵ_2 and γ denote the diffusion and drift coefficient, respectively. In addition we note that upon using the definition (3.2) and functional forms (3.3)–(3.5), integration by parts yields the above drift terms.

3.3.2. Numerical Methods

Here we describe the numerical discretization in space and time for solution of (3.6). We restrict the computations to a small subregion inside the experimental well. Thus, boundary effects due to the finite size of the well are negligible and we may impose periodic boundary conditions at the boundary of the computational domain $\Omega = [0, L_x] \times [0, L_y]$. For the numerical approximation of (3.6) all spatial derivatives are approximated by second-order centered finite differences on a regular grid. The convolution integrals are computed by trapezoidal quadrature, which yields exponential convergence for periodic analytic functions [30]. Hence the numerical discretization error is second-order accurate in space and angle.

For parabolic problems standard explicit Runge-Kutta schemes impose rather stringent restrictions on the time-step for numerical stability, typically $\Delta t \leq C \Delta x^2$, and hence are notoriously inefficient [25]. In contrast, implicit methods waive those time-step restrictions but would require here the solution of a nonlinear integro-differential boundary value problem at every time step, a rather high price to pay.

To avoid the above mentioned difficulties, we opt for Runge-Kutta-Chebyshev methods instead, which are fully explicit while allowing larger time-steps. Instead of maximizing the accuracy, RK-Chebyshev methods maximize the interval $[-\ell, 0]$ of the negative real axis contained in the stability region [25, 33]. Because ℓ is proportional to s^2, for a fixed number of stages, s, any reduction of the mesh size Δx can be counterbalanced by an equivalent increase of the number of stages while keeping the time-step Δt fixed. Therefore RK-Chebyshev methods circumvent the crippling quadratic increase in the number of time-steps of traditional RK methods that results from any linear reduction of the mesh size [3, 22, 51].

For instance, the first-order s-stage RK-Chebyshev method for the initial-value problem

$$y'(t) = f(y), \quad y(0) = y_0, \tag{3.7}$$

is given by

$$g_0 = y_0, \tag{3.8}$$
$$g_1 = y_0 + (1/s^2)\Delta t f(g_0), \tag{3.9}$$
$$g_i = (2/s^2)\Delta t f(g_{i-1}) + 2g_{i-1} - g_{i-2}, \tag{3.10}$$
$$y_1 = g_s. \tag{3.11}$$

In Figure 3.7 we observe that the stability regions of the 3-stage RK-Chebyshev method is about nine times larger than that of the standard fourth-order RK4. Following [24], we eliminate the two intersections where the stability region shrinks to zero by adding small damping of size $\epsilon > 0$. Let $\epsilon > 0$ and $T_s(x)$ denote the Chebyshev polynomial of degree s [1]. Then the damped RK-Chebyshev method for (3.7) is given by

$$g_0 = y_0, \tag{3.12}$$
$$g_1 = y_0 + \Delta t(w_1/w_0)f(g_0), \tag{3.13}$$
$$g_i = \frac{1}{T_i(w_0)}\left[2w_1\Delta t T_{i-1}(w_0)f(g_{i-1}) + 2w_0 T_{i-1}(w_0)g_{i-1} - T_{i-2}(w_0)g_{i-2}\right], \tag{3.14}$$
$$y_1 = g_s. \tag{3.15}$$

where

$$R_s(z) = \frac{1}{T_s(w_0)}T_s(w_0 + w_1 z), \quad w_0 = 1 + \frac{\epsilon}{s^2}, \quad w_1 = \frac{T_s(w_0)}{T_s'(w_0)}. \tag{3.16}$$

As illustrated in Fig. 3.7 for $\epsilon = 0.05$, the stability domain is now slightly shorter (by a factor $4\epsilon s^2/3$), but its boundary remains at a safe distance form the real axis [24].

When the right-hand side in (3.7) explicitly depends on time, the terms involving $f(g_i)$ in (3.12)–(3.15) are replaced by $f(g_i, t_i)$. The precise times $t_i \in [0, \Delta t]$ where f needs to be evaluated are determined by augmenting (3.7) with the trivial differential equation,

$$z'(t) = 1, \quad z(0) = t_0 \tag{3.17}$$

and applying (3.12)–(3.15) to it. Thus for $t \in [0, \Delta t]$ we have

$$t_0 = 0, \tag{3.18}$$
$$t_1 = \Delta t(w_1/w_0), \tag{3.19}$$
$$t_i = \frac{1}{T_i(w_0)}\left[2w_1\Delta t T_{i-1}(w_0)) + 2w_0 T_{i-1}(w_0)t_{i-1} - T_{i-2}(w_0)t_{i-2}\right], \tag{3.20}$$

and so forth during subsequent time steps.

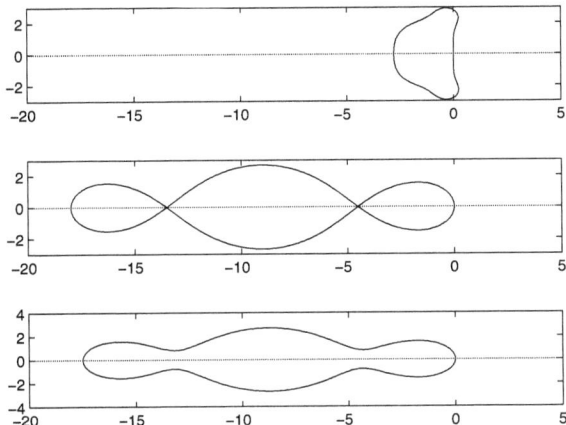

Figure 3.7.: *Stability regions: fourth-order RK4 (top) and first-order 3-stage Chebyshev methods without damping (middle) and with damping (bottom). The stability regions of the RK-Chebyshev method is about nine times larger then that of the standard RK4.*

3.4. Stability

Linear stability analysis

Before investigating numerically the emergence of coherent patterns in the full nonlinear model (3.6), it is instructive to investigate the stability properties of the homogeneous state, i.e. the state, where the density of cells in angular and spatial space is equally distributed. Linear stability analysis characterizes the effect of small perturbations on the early time evolution in angular and spatial space. We therefore expect good agreement with the early stages of the numerical solution of the fully nonlinear model (3.6).

If $\rho = 0$, the homogeneous state $C \equiv \bar{C}$ is an exact solution of (3.6), since we then have $W \star C = C$. In this case, linearization of (3.6) about $C \equiv \bar{C}$ using the ansatz

$$C(\vec{x}, \theta, t) = \bar{C} + \delta\, C'_{n,q}(\vec{x}, \theta, t), \tag{3.21}$$

where the amplitude of the perturbation $\delta \ll 1$ is small, results in an eigenvalue problem for the integro-differential operator previously analyzed in [16, 35]. In particular, Mogilner et al. [35] showed for an unbounded spatial domain that the functions

$$u_{q_1}(x) = e^{iq_1 x}, \quad u_{q_2}(y) = e^{iq_2 y}, \quad z_n(\theta) = e^{in}, \qquad q_1, q_2 \in \mathbb{R} \tag{3.22}$$

form a complete set of orthogonal eigenfunctions for the spatial and angular Laplace operator with eigenvalues q_1, q_2 for the spatial and $n = 0, 1, 2, \ldots$ for the angular diffusion operator, respectively. In addition, they proved that (3.22) are not only the eigenfunctions of the Laplace operators, but also of the convolution operators W_1* and W_2*, where the eigenvalues are the Fourier coefficients denoted by $\hat{W}_n$ and $\hat{W}_q = \hat{W}_{q_1}\hat{W}_{q_2}$, where $q = \sqrt{q_1^2 + q_2^2}$ and

$$\hat{W}_n = \frac{1}{\pi}\int_{-\pi}^{\pi} W_1(\theta) e^{-in\theta}\, d\theta, \qquad \hat{W}_{q_j} = \int_0^{L_{x_j}} W_2(x_j) e^{-iq_j x_j}\, dx_j, \qquad (3.23)$$

where $j = 1, 2$ and $x_1 = x$, $x_2 = y$. From the normal modes ansatz

$$C'_{n,q}(\vec{x}, \theta, t) = e^{\lambda t}\, u_{q_1}(x)\, u_{q_2}(y)\, z_n(\theta) \qquad (3.24)$$

the stability of the homogeneous state is then found from the solution of the equation

$$\lambda = -r + \overline{C}\, s, \qquad (3.25)$$

where

$$r = (\epsilon_1 n^2 + \epsilon_2 q^2) \quad \text{and} \quad s = \hat{W}_q \hat{W}_n \gamma (n^2 + q^2) \qquad (3.26)$$

for all q_1, q_2 and n. Hence the condition for instability of the homogeneous solution is given by $\lambda > 0$. Thus, any increase in the diffusion coefficients ϵ_1, ϵ_2 tends to stabilize the system, while the cell-to-cell interaction terms $\hat{W}_n$ and $\hat{W}_q$ tend to destabilize the system, for increasing values of n, q, unless $\hat{W}_n$ is zero. Moreover, for any particular values of ϵ_1, ϵ_2, n, q, the constant state $\overline{C}$ becomes unstable at sufficiently high cell density, unless $\hat{W}_n$ or $\hat{W}_q$ vanishes.

For our extended model with logistic growth, where $\rho = 0$, the homogeneous state about which we linearize is now time-dependent, due to the slow mass increase. Thus we make the ansatz

$$C(\theta, \vec{x}, t) = \overline{C}(t) + \delta\, C'_{n,q}(\theta, \vec{x}, t) \qquad (3.27)$$

with $C'_{n,q} = \hat{C}_{n,q}(t) e^{i(qx+n\theta)}$, since now the standard normal modes ansatz may lead to non-normal linear systems with non-orthogonal eigenfunctions – see [47], for instance, for a more detailed discussion of such problems in the context of hydrodynamics. Our slightly more general ansatz for $C(\theta, \vec{x}, t)$ then leads to the following differential equation for $\hat{C}_{n,q}$

$$\frac{d\hat{C}_{n,q}}{dt} = \left[-(\epsilon_1 n^2 + \epsilon_2 q^2) + \overline{C}(t)\hat{W}_q \hat{W}_n \gamma(n^2 + q^2) + \rho\left(1 - \frac{\overline{C}(t)}{\kappa}\right) \right] \hat{C}_{n,q}(t), \qquad (3.28)$$

where

$$\overline{C}(t) = \frac{\kappa}{1 + C_1 \kappa e^{-t}} \qquad (3.29)$$

is the solution of the leading order problem and represents the slowly growing mass until the carrying capacity is reached. The constant $C_1 = 1/\overline{C}(0) - 1/\kappa$, where $\overline{C}(0)$ is chosen to be the same as $\overline{C}$ in the original problem and we denote $\kappa = K/(2\pi L_x L_y)$. Hence, the growth rate is given here by

$$\ln(\hat{C}_{n,q}(t)) = (-r + s\kappa)t + \frac{s\kappa - \rho}{\rho}\ln\left(1 + C_1 \kappa\, e^{-t}\right) + const. \qquad (3.30)$$

We note that now the additional parameter K, the carrying capacity, will have a decisive impact on the stability properties of the solutions.

Comparison of the full model with linear stability
To compare the results from linear stability analysis to those from the numerical simulation of the full problem, we choose as an example the simple case for which $\rho = 0$ and $\gamma = 1$. We set the (constant)

base state $\bar{C} = 25$, let $\epsilon_1 = 0.0025$ and $\epsilon_2 = 0.5$ so that for $n > 0$ and $q = 0$ the base state is unstable according to linear stability analysis. Now, we determine n_{max} such that the growth rate is maximal, i.e. $\sigma_{max} = \sigma(n_{max}, q_{max})$ (here $q = q_{max} = 0$). Thus, we can find n_{max} which is at most $O(1)$ with a σ_{max} not too small, together with the the corresponding eigenfunction $C'_{n,q}$ and a corresponding asymptotic growth rate λ.

Next, we initialize our nonlinear simulation with the initial data

$$\bar{C} + \delta C'_{n,q}, \quad \text{such that} \quad \delta \leq \min\left(0.1, \, 0.1 \frac{\sigma_{max}}{n_{max}^2}\right)$$

to ensure that the correction term does not invalidate the original assumptions of linear stability analysis.

In figure 3.8, $\log \|C\|$ with

$$\|C\| = \frac{\max_{x,y} |C(\theta, x, y, t) - \bar{C}|}{\delta}$$

is shown versus t, both for the solution of the fully nonlinear model and for that from linear stability. Note that the growth rate of the linearized problem for the extended model, i.e. where $\rho = 0$, now also depends on time. Once initial transients have died out, both models agree, as expected. As time progresses, however, the dynamics of the full model deviate from those of the linearized problem. Thus, the evolving patterns may deviate from those predicted by linear stability theory, in particular at later times, as the cell culture reaches confluence, depending on parameter values.

In figure 3.8 we show a comparison of the growth rates for the fully nonlinear and the linearized models, with the above set of parameters. This example illustrates the following generic behaviour. For $\rho = 0$, we observe agreement right from the beginning, since the perturbation corresponds to an exact eigenfunctions, as in the linear stability problem. For the extended model with $\rho = 0.2$ and $K = 1220$, for instance, we observe that the long-time behaviour of the solution of (3.30) compares well with the solution to the full problem (3.6). Eventually though, the nonlinear terms come into play and the solution of the full model deviates from the prediction of the linear model.

3.5. Comparison of simulations with experiments

3.5.1. Parameter values

To compare computational results from any mathematical model with those from experiment, it is crucial to have accurate estimates of the parameter values. While the values of most parameters were determined quite accurately from experiment, uncertainties about some of them remained.

In particular, the estimation of the spatial diusion coe cient ϵ_2 was performed by hand following the positions of the moving cells. From the experiment (table C.1), the average spatial diusion coe cient $\epsilon_2 = 0.29 \, \mu m^2/s = 0.025$ mm^2/days. Because cells do not change their orientation in a continuous way, ϵ_1 could not be determined from those measurements, mainly because cells also contract, become almost spherical, and elongate randomly, excluding the determination of a well-defined axis of orientation. In our numerical simulations, we chose values for the angular diusion coe cient ϵ_1, which is the mean square displacement in angular space per day, to range from 0.025 (Figs. 3.9, 3.10, 3.12) to 0.0025 (Fig. 3.11). The values for ρ and K are always given by 1.2 and 4000 cells/mm^2 and have been determined previously in [6]. They were used as initial guess for a nonlinear least-squares parameter fit to the time evolution of the total mass. The size of the domain $L_x = 3.75$mm and $L_y = 2,75$mm was chosen to match the area observable under the microscope. In [18], the critical angle α was obtained for fibroblast cultures by inspection of relative angles between cells at confluence. Because of the strong similarity between cytokine cultured chondrocytes and fibroblasts, we used the same value here, i.e. $\alpha = 20^0$ for our simulations, except in Fig. 3.12, where we also include results for critical angles $\alpha = 40^0$ and 60^0. Since chondrocytes only attach when they are very close to each other, we chose the standard deviation $\sigma = 0.01$mm for the spatial interaction kernel to be about the length of a single cell. The value of $\gamma = 0.0005$ essentially sets the convective time scale and was obtained by fitting the cluster size from the simulation to that obtained from experiment – see Section 2.3.

3.5.2. Numerical simulations

Starting from a random initial distribution at $t = 0$, we solve (3.6) using the numerical method described in section 3.3.2 and the parameter values listed above. In Fig. 3.9, snapshots of the cell density at dierent times are shown. Here at each point $(x, y) \in$ the marginal spatial cell density of C, that is the integral of $C(x, y, \theta, t)$ over θ, is displayed. The color used at any point (x, y) corresponds to the angle, where $C(x, y, \theta, t)$ is maximal; hence, it represents the local dominant orientation of the cells. We observe that the number of cells increases uniformly throughout the computational domain , yet past day 6 several patches of cells with a common orientation emerge and settle in a stationary configuration by day 9; note that the total number of cells hardly changes beyond day 6 anymore.

In Fig. 3.10 we compare the simulation with the experimental data using Gabor filters for post-processing both – see section 3.2.3. In doing so the spatial resolution of the microscope image was coarsened to match that of the simulation, while the angular dependence over $[0, \pi)$ was divided into four classes, that is sub-intervals of identical lengths, each one assigned with a dierent color. The cluster size (intersection of $r(s)$ with the x-axis, see Section 3.2.3) was calculated for three samples from the same donor. By fitting the average cluster size from the simulation to that from experiment, between 15 and 20 pixels or about $0.5mm$, we determined the standard value of γ, as shown in Fig. 3.10.

Once the model has been validated through comparison to experiment, it is instructive to change the value of individual parameters to study their eect on the si ze and shape of the patterns at confluence. Thus we can also evaluate the parameter sensitivity of the model and address the uncertainties associated with some of the values obtained from experiment. For instance, the reduction of the angular diusion coe cient ϵ_1 has little eect on the size of the patterns, but the interfac es appear more well-defined in contrast to the standard case: compare Fig. 3.10 and Fig. 3.11. An increase in the critical angle α instead, results in larger and increasingly irregular patterns, while the uniform spatial population density is maintained, as shown in Fig. 3.12

3.6. Concluding remarks

Starting from the classical models by Mogilner et al. [35], we have developed a mathematical model for proliferating chondrocytes, cultured with specific growth factors, by including logistic growth and studied the patterns emerging at confluence through experiments and simulation. Most parameters in the model were obtained directly through independent experiments or from our previous micro-colony tests [6]. Guided by these parameter studies we arrived at reasonable parameter values for comparison to the experimentally observed cell patterns at confluence. Linear stability analysis was used as guidance through the range of unstable parameter values, as their interplay leads to pattern formation; their improved understanding and control will be useful in the future design of engineered tissue.

For the time integration of the nonlinear integro-partial dierential equation, we opted for Runge-Kutta-Chebyshev methods which permit much larger time steps than standard Runge-Kutta methods while nonetheless remaining fully explicit. Quantitative comparison of experimental data with the numerical simulations was achieved in two steps. First, we vizualized the orientation and alignment of the cells with Gabor filters. Then, we determined the average cluster size of the cell population both in the simulation and the experiment.

From the stability analysis and the simulations, we were able to determine key parameters for pattern formation. In particular, we find that the total number of dominant directions of alignment in a cell culture is mainly regulated by the critical angle, below which the probability that cells align is high. Indeed, smaller values in the critical angle α for cell-cell interactions lead to arrays of aligned cells, as observed in experiments, whereas larger values lead to a single dominant direction of alignment. Regarding the diffusion and drift coefficients, both tend to destabilize the homogeneous state and thereby lead to pattern formation. For fixed diffusion coefficients, the average pattern size typically scales with the drift coefficient, γ, although the number of dominant directions of alignment remains identical, as it is regulated by the critical angle.

While we think that continuum models in combination with some local experimental analysis yields convincing evidence to capture the large scale long-time structures of proliferating cell cultures, our work also leaves a number of open tasks and questions. Apart from the study of aggregation patterns, which has been left open, the experimental determination of the remaining parameters, in particular drift parameters but also angular diffusion coefficients will be an important future task. We believe that in principle more sophisticated image analysis and segmentation software would allow the automatic tracking of larger number of cells and yield both more refined and improved statistics. Through a new set of experiments, more experimental studies such as those by Elsdale [17] are needed in order to establish more accurately the critical angle for cell alignment for the particular cells under consideration without relying on similar cases from the literature.

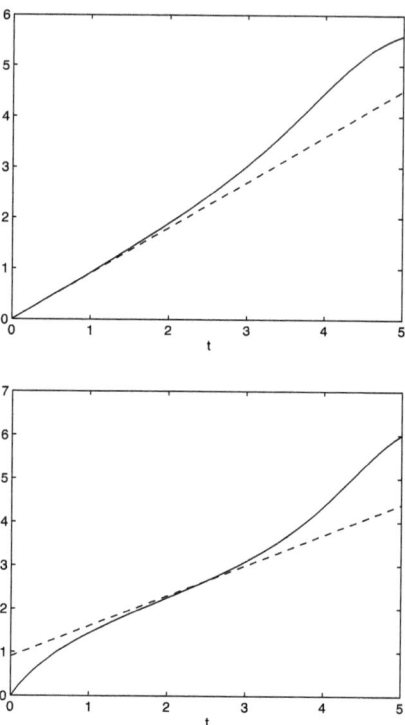

Figure 3.8.: *Comparison of the growth rates for the fully nonlinear and the linearized models: Left, $= 0$: The dashed line is t, where is given by the solution of the linear stability problem (3.25). The solid curve denotes $\ln(||C||)$. Right, $= 0.2$: The dashed line shows the long-time behaviour of the solution of (3.30). The solid curve results from the solution to the full problem (3.6).*

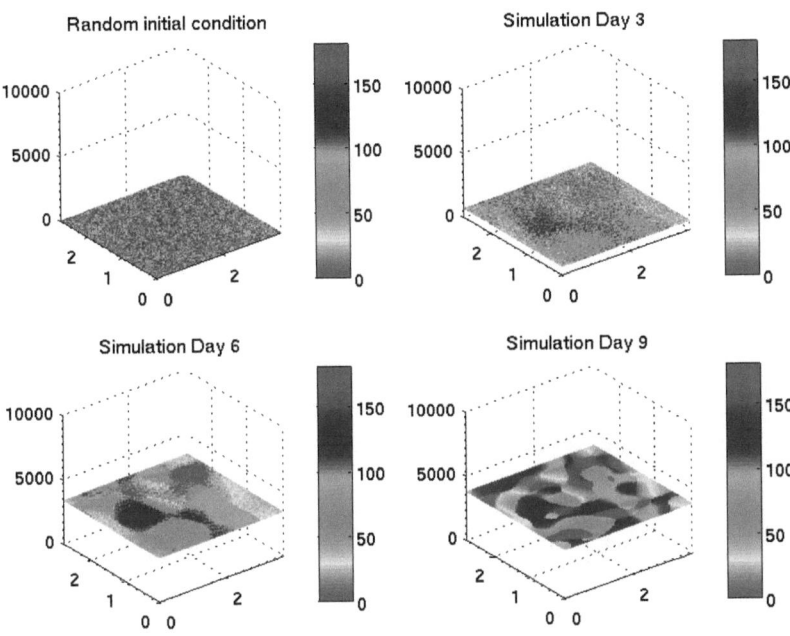

Figure 3.9.: *Snapshots of the cell density at different times. At each point (x, y) in the computational domain, the marginal angular density of C is shown, that is the number of cells at position (x, y) and time t; the color represents the angle for which $C(x, y, , t\ )$ is maximal.*

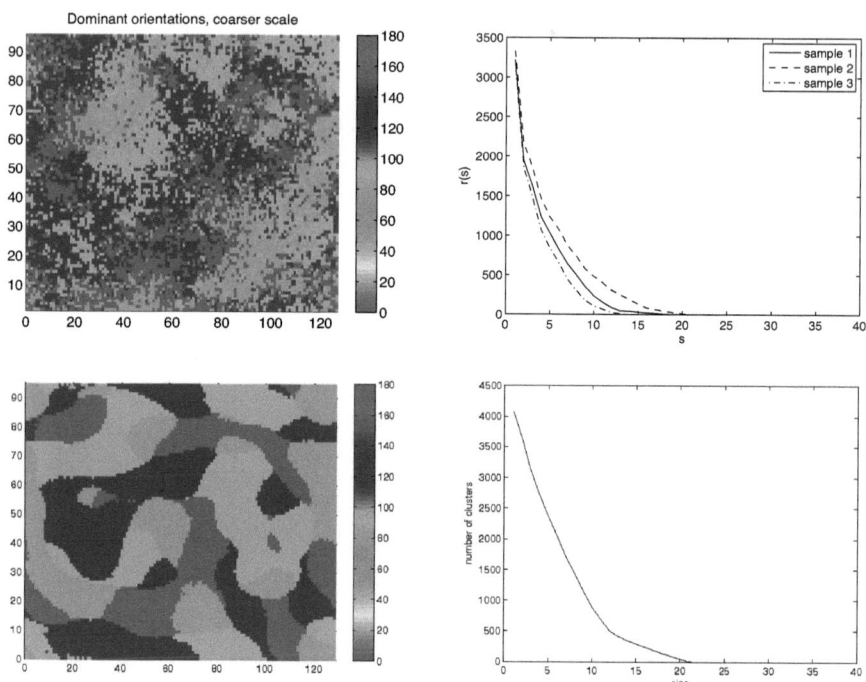

Figure 3.10.: *Comparison of simulation (bottom) with experiment (top). Here color indicates the local dominant orientation of the cells. The cluster size for three samples from the same donor (top) and for the simulation (bottom) are shown on the right.*

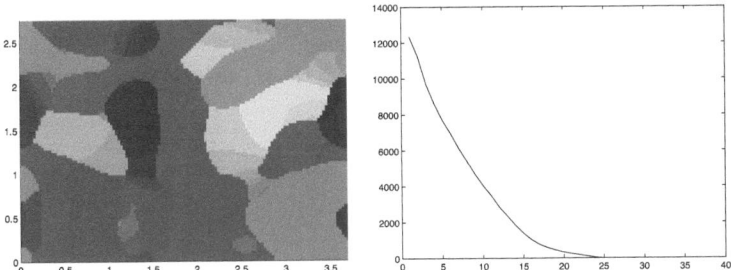

Figure 3.11.: *The cell density is shown at confluence for the smaller angular diffusion coefficient:* $\nu_1 = 0.0025$. *Left: the dominant cell orientation; right: the cluster size function* $r(s)$.

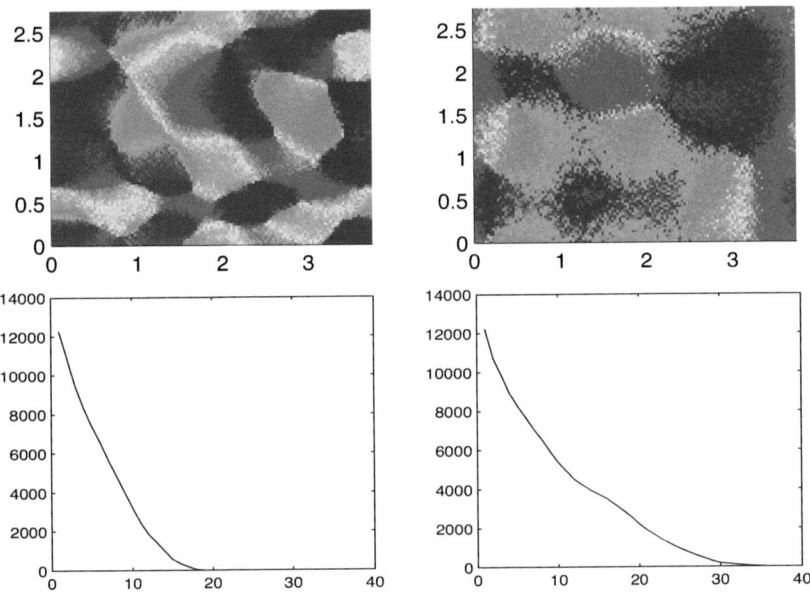

Figure 3.12.: *The cell density is shown at confluence for larger critical angles:* $\alpha = 40^0$ *(left) and* $\alpha = 60^0$ *(right). The corresponding cluster size functions* $r(s)$ *are shown below.*

4. Analysis of the spatio-angular model

In this chapter we would like to analyze in details our model from a theoretical point of view. In particular we want to show the existence of a weak solution under reasonable assumptions in the first section and explain the linear stability analysis in the second.

4.1. Existence of a weak solution

Our domain $U = (-\pi, \pi) \times D$, where $D = [0, L_x] \times [0, L_y]$, is an open and bounded subset of $\mathbb{R}^3$. We set $U_T = U \times (0, T]$ for some fixed time $T > 0$. The equation for the variation of the cell density $C(\theta, x, y, t)$ ($\theta \in (-\pi, \pi)$, $(x, y) \in D$, $t \in (0, T]$) takes in account the random movement, the doubling of the cells and their interactions. The equation is deeply described in chapter 3 and has the following form

$$\begin{aligned}
\frac{\partial C}{\partial t} &= \epsilon_1 \frac{\partial^2 C}{\partial \theta^2} + \epsilon_2 \left(\frac{\partial^2 C}{\partial x^2} + \frac{\partial^2 C}{\partial y^2} \right) + \rho C \left(1 - \frac{K}{L_x L_y} \int_{-} C \right) - \gamma_1 \frac{\partial}{\partial \theta} \left(C \left[\frac{\partial W}{\partial \theta} * C \right] \right) \\
&\quad - \gamma_2 \left\{ \frac{\partial}{\partial x} \left(C \left[\frac{\partial W}{\partial x} * C \right] \right) + \frac{\partial}{\partial y} \left(C \left[\frac{\partial W}{\partial y} * C \right] \right) \right\}.
\end{aligned} \quad (4.1)$$

First we nondimensionalize the governing equations (4.1) as follows, with:

$$C^* = \frac{C}{K} L_x L_y, \quad t^* = \rho t, \quad \theta^* = \theta \sqrt{\frac{\rho}{\epsilon_1}}, \quad x^* = x \sqrt{\frac{\rho}{\epsilon_2}}, \quad y^* = y \sqrt{\frac{\rho}{\epsilon_2}}.$$

and

$$\alpha^* = \alpha \sqrt{\frac{\rho}{\epsilon_1}}, \quad \sigma^* = \sigma \sqrt{\frac{\rho}{\epsilon_2}}.$$

Then

$$C = \frac{K}{L_x L_y} C^*, \quad L^* = L \sqrt{\frac{\rho}{\epsilon_1}}, \quad L_x^* = L_x \sqrt{\frac{\rho}{\epsilon_2}}, \quad L_y^* = L_y \sqrt{\frac{\rho}{\epsilon_2}}.$$

Calculating the derivative we have

$$\frac{\partial C}{\partial t} = \frac{\partial (C^* \frac{K}{L_x L_y}) \, dt^*}{dt^* \, dt} = \frac{K \rho}{L_x L_y} \frac{\partial C^*}{\partial t^*},$$

$$\frac{\partial^2 C}{\partial \theta^2} = \frac{\partial^2 (C^* \frac{K}{L_x L_y})}{\partial \theta^{*2}} \left(\frac{d\theta^*}{d\theta} \right)^2 = \frac{K \rho}{L_x L_y \epsilon_1} \frac{\partial^2 C^*}{\partial \theta^{*2}},$$

and similarly for x and y.
Dividing by $K\rho/(L_x L_y)$ and dropping the '*' for simplicity of notation, we thus obtain the nondimensional form

$$\begin{cases} \partial_t C = \text{div}(\nabla C) - A \, \text{div}(C \, W * C) + f(C) & \text{in } U_T \\ C = g & \text{on } U \times \{t = 0\}, \end{cases} \quad (4.2)$$

with initial condition $g : U \to \mathbb{R}$, $f(C) = C\left(1 - \int_- C d\theta\right)$, and $A = K\gamma/(L_x L_y \epsilon)$; without loss of generality we assume $\epsilon_1 = \epsilon_2 = \epsilon$ here.

Let's introduce the variable $z = (x, y, \theta) \in U$. We temporarily suppose that $C = C(z, t)$ is in fact a smooth solution of our problem and switch our viewpoint, by associating with C a mapping which for simplicity we call with the same name

$$C : [0, T] \to H^1_{per}(U)$$

defined by

$$[C(t)](z) := C(z, t) \quad (z \in U, 0 \le t \le T),$$

where we remember that $H^1_{per}(U)$ is the Sobolev space

$$H^1_{per}(U) = \{u : U \to \mathbb{R} | u \in L^2(U), \ \nabla u \in L^2(U)\},$$
$$L^p(U) = \{u : U \to \mathbb{R} | u \text{ is Lebesgue measurable}, \ \|u\|_{L^p(U)} < \infty \},$$

with the norms

$$\|u\|_{H^1_{per}(U)} = \|u\|_{L^2(U)} + \|\nabla u\|_{L^2(U)},$$

$$\|u\|_{L^p(U)} = \begin{cases} \left(\int_U |u|^p dx\right)^{\frac{1}{p}} & \text{if } 1 \le p < \infty \\ \operatorname{ess\,sup}_U |u| & \text{if } p = \infty. \end{cases}$$

In other words, we are going to consider C not as a function of z and t together, but rather as a mapping C of t into the space $H^1_{per}(U)$ of functions of z. Then if we fix a function $v \in H^1_{per}(U)$, we can multiply equation (4.2) by v and integrate by parts, to find

$$(C', v) + (\nabla C, \nabla v) = A(C - W * C, v) + (f(C), v) \tag{4.3}$$

for each $0 \le t \le T$, where $(,)$ denotes the inner product in $L^2(U)$

$$(u, v) = \int_U u\, v\, dx.$$

In fact, the integration by parts formula for each $u, v \in H^1_{per}(U)$ can be written as

$$\int_U \nabla u\, v\, dx = -\int_U u\, \nabla v + \int_{\partial U} u\, v\, \nu dS,$$

where ν is the outer normal vector. We observe that the last term vanishes for periodic conditions on the boundary.

We consider the dual space of $H^1(U)$, $H^{-1}(U)$ with norm

$$\|u\|_{H^{-1}(U)} = \sup\{<u, v> | v \in H^1_{per}(U), \ \|v\|_{H^1_{per}(U)} \le 1\},$$

denoting with $<,>$ the pairing between $H^{-1}(U)$ and $H^1_{per}(U)$ and introduce for any Banach space X the space

$$L^2(0, T; X) = \{u : [0, T] \to X, \text{ measurable, with } \|u\|_{L^2(0,T;X)} < \infty \}$$

where

$$\|u\|_{L^2(0,T;X)} = \left(\int_0^T \|u(t)\|_X^2 dt\right)^{\frac{1}{2}}.$$

Our goal is to find a weak solution of (4.2), that means a function

$$C \in H^1(0, T; H^1_{per}(U)), \quad \text{with } C' \in L^2(0, T; H^{-1}(U)),$$

which solves the problem

$$<C', v> + (\nabla C, \nabla v) = A(C \nabla W * C, \nabla v) + (f(C), v) \quad (4.4)$$

for each $v \in H^1_{per}(U)$, $0 \leq t \leq T$ and $C(0) = g$. For the regularity of the solution all the terms of this equation are well-defined. For the time and space dependency, we need $C \in H^1 \subset L^4(U)$ to be sure that all products are in $L^2(U)$. Indeed if $\nabla W \in L^\infty(U)$ then $\nabla W * C \in L^2(U)$ which implies, for $\nabla C \in L^2(U)$, $\nabla C \nabla W * C \in L^2(U)$. For the function $f(C)$ it suffices to observe that integrating the function C only in one variable, we obtain again a function in $L^4(U)$ for the Fubini's theorem.

4.1.1. Maximum principle and mass control

Theorem 4.1.1. *Let C be a smooth and bounded solution of equation (4.2). If $C(x, 0) \geq 0$ $\forall x \in U$, then $C(x, t) \geq 0$ $\forall (x, t) \in U_T$ and the mass is bounded, i.e.*

$$\|C\|_{L^1(U)} \leq K_1 := e^T \|C(0)\|_{L^1(U)}. \quad (4.5)$$

Proof. Let define the negative part of C, $C_- = \min(C, 0)$ and test equation (4.2) with it. Because the term at the boundary vanish, we have

$$(C', C_-) + (\nabla C, \nabla C_-) = A(C \nabla W * C, \nabla C_-) + (f(C), C_-). \quad (4.6)$$

From a theorem of the function analysis we have that if $C \in H^1_{per}(U)$ then $C_- \in H^1_{per}(U)$; moreover $C' = C'_-$ and $\nabla C_- = \nabla C$ if C is negative. Using this fact, we observe that $(C', C_-) = (C_-', C_-)$, $(\nabla C, \nabla C_-) = (\nabla C_-, \nabla C_-)$ and $(f(C), C_-) = (f(C_-), C_-)$; indeed for C positive it is trivial because its negative part is zero and for C negative, $C = C_-$ and the derivatives remain the same.

$$(C_-', C_-) + (\nabla C_-, \nabla C_-) = A(C_- \nabla W * C, \nabla C_-) + (f(C_-), C_-) \quad (4.7)$$

Now, using (see [20])

$$(C_-', C_-) = \frac{1}{2}\frac{d}{dt}\int_U |C_-|^2, \quad (4.8)$$

our equation can be reduced to

$$\frac{1}{2}\frac{d}{dt}\int_U |C_-|^2(t) + \int_U |\nabla C_-|^2 = A\int_U C_- \nabla W * C \nabla C_- + \int_U f(C_-)C_-. \quad (4.9)$$

Writing down the expression of f, we obtain

$$\frac{1}{2}\frac{d}{dt}\int_U |C_-|^2(t) + \int_U |\nabla C_-|^2 = A\int_U C_- \nabla W * C \nabla C_- + \int_U |C_-|^2 - \int_U \int_- C_- d\theta |C_-|^2. \quad (4.10)$$

Moreover the θ-average of C_- is bounded by a constant hypothesis, then exist a constant B such that

$$\frac{1}{2}\frac{d}{dt}\int_U |C_-|^2(t) + \int_U |\nabla C_-|^2 \leq A\int_U C_- \nabla W * C \nabla C_- + \int_U |C_-|^2 + B\int_U \int_- |C_-|^2. \quad (4.11)$$

Sinnce $\nabla W \in L^\infty(U)$ and C is bounded we have $\nabla W * C \in L^\infty(U)$, then $\|\nabla W * C\|_{L^\infty(U)} \leq K$ for a constant $K > 0$. Then applying the Cauchy inequality

$$\int_U C_- \nabla W * C \nabla C_- \leq \|C_-\|_{L^2(U)} \|\nabla W * C\|_{L^\infty(U)} \|\nabla C_-\|_{L^2(U)} \leq$$
$$K\|C_-\|_{L^2(U)} \|\nabla C_-\|_{L^2(U)} \leq K\left(\frac{1}{4\delta}\|C_-\|^2_{L^2(U)} + \delta\|\nabla C_-\|^2_{L^2(U)}\right). \quad (4.12)$$

For $\delta = 1/(AK)$ the norm of the gradient vanish and we finally have

$$\frac{d}{dt} \|C_-\|^2_{L^2(U)} \leq \frac{A^2K^2 + 4 + 4B}{2} \|C_-\|^2_{L^2(U)}. \tag{4.13}$$

With the Gronwall inequality and calling $\eta(t) = \|C_-(t)\|^2_{L^2(U)}$ we obtain

$$\eta(t) \leq e^D \eta(0), \tag{4.14}$$

for $D = \frac{A^2K^2+4+4B}{2}$. But $\eta(0) = 0$ for the choice of the initial conditions. Then is $\eta(t) = 0$, implying that $C_- = 0$ a.e. in U and for all $0 < t < t_1$. Now, if we take as initial condition $C(x, t_1)$, we can find a time $t_1 < t_2 < T$ in which C is positive, and so on until T, which proves the first part of th theorem. If we integrate equation (4.2) over the entire domain and applying the Gauss theorem, we obtain

$$\frac{\partial}{\partial t} \int_U C = \int_{\partial U} C \cdot \vec{n} - A \int_{\partial U} C \ W * C \cdot \vec{n} + \int_U f(C), \tag{4.15}$$

where $\vec{n}$ the unit outward normal defined at points of ∂U, normal. For the periodicity of the solution the integrals on the boundary vanish. Since for theorem (4.1.1) the function C is positive under our assumptions, we have

$$\frac{\partial}{\partial t} \int_U C(t) = \int_U f(C(t)) = \int_U C(t) - \int_U C(t) \int_- C(t) \leq \int_U C(t). \tag{4.16}$$

We can now apply the Gronwall theorem to the function $\gamma(t) = \int_U C$, to obtain

$$\int_U C(t) \leq e^t \gamma(0) \leq e^T \gamma(0) = e^T \|C(0)\|_{L^1(U)}. \tag{4.17}$$

$\square$

4.1.2. Galerkin approximations

To prove the existence of a weak solution we use the Galerkin's method which consists in constructing solutions of certain finite-dimensional approximations and then passing to limits. Assume the functions $w_k = w_k(x)$, $k = 1, \ldots$ are smooth, $\{w_k\}_{k=1}^\infty$ is an orthogonal basis of $H^1_{per}(U)$ and an orthonormal basis of $L^2(U)$.

Theorem 4.1.2. For each integer $m = 1, 2, \ldots$ there exists a unique function $C_m : [0, T] \to H^1_{per}(U)$ of the form

$$C_m(t) := \sum_{l=1}^m d_m^l(t) w_l, \tag{4.18}$$

with

$$d_m^l(0) = (g, w_l), \quad l = 1, \ldots, m, \tag{4.19}$$

which solves the problem

$$(C_m', w_k) + (\ C_m, \ w_k) = A\ (C_m \ W * C_m, \ w_k) + \int_U f(C_m) w_k \tag{4.20}$$

for $0 \leq t \leq T$, $k = 1, \ldots, m$.

Proof. Indeed, substituting (4.18) in (4.20) we obtain

$$(C'_m, w_k) = \left(\left(\sum_{l=1}^m d^l_m(t)w_l\right)', w_k\right) = \sum_{l=1}^m {d^l_m}'(t)(w_l, w_k) = {d^k_m}'(t),$$

for the first term. The other three terms are

$$(\triangle C_m, w_k) = \left(\triangle \sum_{l=1}^m d^l_m(t)w_l, w_k\right) = \sum_{l=1}^m (\triangle w_l, w_k) d^l_m(t),$$

$$(C_m \ W * C_m, w_k) = \left(\sum_{l=1}^m d^l_m(t)w_l \ W * \sum_{l=1}^m d^l_m(t)w_l, w_k\right)$$

$$= \left(\sum_{l=1}^m d^l_m(t)w_l \sum_{l=1}^m d^l_m(t) \ W * w_l, w_k\right),$$

$$\int_U f(C_m)\, w_k = \int_U f\left(\sum_{l=1}^m d^l_m(t)w_l\right) w_k. \tag{4.21}$$

For fixed k these terms are real numbers that depend locally Lipschitz continuously on $(d^1_m(t),\dots,d^m_m(t))$. Then (4.20) is an ordinary differential equation for this vector, subject to the initial conditions (4.19). According to the standard existence theory for ordinary differential equations, there exists a unique absolutely continuous function $(d^1_m(t),\dots,d^m_m(t))$ satisfying (4.19) and (4.20) for a.e. $0 \leq t \leq T$, at least for small T. And then C_m defined by (4.18) solves (4.20) for a.e. $0 \leq t \leq T$. □

Observation 4.1.3. *Because of theorem 4.1.1 it is reasonable to make the assumption that the functions C_m are nonnegative and with bounded mass. To proof this statement could be long and complicate and it is not our goal here.*

4.1.3. A-priori estimates

Theorem 4.1.4. *Under assumptions (4.1.3), there exists a constant K, depending only on U, T, such that*

$$\max_{0 \leq t \leq T} \|C_m(t)\|_{L^2(U)} + \|C_m\|_{L^2(0,T;H^1_{per}(U))} + \|C'_m\|_{L^2(0,T;H^{-1}(U))} \leq K\, \|g\|_{L^2(U)}. \tag{4.22}$$

Proof. We shall estimate every term on the left side.

1. We multiply equation (4.20) by $d^m_k(t)$, sum for $k = 1,\dots,m$ to find

$$(C'_m, C_m) + (\triangle C_m, C_m) = A\,(C_m \ W * C_m, C_m) + \int_U f(C_m)\, C_m,$$

for a.e. $0 \leq t \leq T$. Then with the Hölder inequality

$$\frac{d}{dt}\left(\frac{1}{2}\|C_m\|^2_{L^2(U)}\right) + \|\nabla C_m\|^2_{L^2(U)} \leq A\, \|C_m\|_{L^2(U)} \|C_m\|_{L^2(U)} + \|C_m\|^2_{L^2(U)}.$$

In fact

$$(C_m \ W * C_m, C_m) \leq \|W * C_m\|_{L^\infty(U)} \int_U C_m\, C_m$$

$$\|W * C_m\|_{L^\infty(U)} = \operatorname*{ess\,sup}_U |W * C| = \operatorname*{ess\,sup}_U \left|\int_U W(x - x')C(x')dx'\right|$$

$$\leq \operatorname*{ess\,sup}_U \int_U |W(x - x')||C(x')| \leq \|W\|_{L^\infty(U)} \|C_m\|_{L^1(U)} \leq K_1,$$

for the choice of the kernel W and theorem (4.1.1). For the boundness of C we have that $\int_{pi} C$ is bounded, then using the Cauchy inequality, we obtain

$$\frac{d}{dt}\|C_m\|^2_{L^2(U)} + 2\|\nabla C_m\|^2_{L^2(U)} \leq 2AK_1\left(\delta \|\nabla C_m\|^2_{L^2(U)} + \frac{1}{4\delta}\|C_m\|^2_{L^2(U)}\right) + 2\|C_m\|^2.$$

Choosing $\delta = 1/(AK_1)$, $K_2 = (A^2K^2 + 4)/2$, we have

$$\frac{d}{dt}\|C_m\|^2_{L^2(U)} \leq K_2\|C_m\|^2_{L^2(U)}.$$

If we call $\eta(t) = \|C_m(t)\|^2_{L^2(U)}$, with the Gronwall inequality we obtain

$$\eta(t) \leq e^{K_2 t}\eta(0).$$

Being
$$\eta(0) = \|C_m(0)\|^2_{L^2(U)} \leq \|g\|^2_{L^2(u)},$$

for $K = e^{K_2 T}$, we have the first estimate

$$\max_{0 \leq t \leq T}\|C_m(t)\|^2_{L^2(U)} \leq K\|g\|^2_{L^2(U)}.$$

2. Integrating the last equation from 0 to T, we get automatically the second estimate

$$\|C_m(t)\|^2_{L^2(0,T;H^1_{per}(U))} = \int_0^T \|C_m\|^2_{H^1_{per}(U)} dt \leq KT\|g\|^2_{L^2(U)}.$$

3. Fix any $v \in H^1_{per}(U)$, with $\|v\|_{H^1_{per}(U)} \leq 1$, and write $v = v^1 + v^2$, where $v^1 \in \text{span}\{w_k\}_{k=1}^m$ and $(v^2, w_k) = 0$, $(k = 1,\ldots,m)$. Since the functions $\{w_k\}_{k=0}^\infty$ are orthogonal in $H^1_{per}(U)$, $\|v^1\|_{H^1_{per}(U)} \leq \|v\|_{H^1_{per}(U)} \leq 1$. Utilizing 4.20, we deduce for a.e. $0 \leq t \leq T$ that

$$(C'_m, v^1) + (\nabla C_m, \nabla v^1) = A(\nabla C_m \cdot W * C_m, v^1) + \int_U f(C_m)v^1.$$

As $(C'_m, v^1) = (C'_m, v)$ we have

$$(C'_m, v) = -(\nabla C_m, \nabla v^1) + A(\nabla C_m \cdot W * C_m, v^1) + \int_U f(C_m)v^1.$$

Since $\|v^1\|_{H^1_{per}(U)} \leq 1$ we obtain

$$|<C'_m, v>| \leq \|\nabla C_m\|_{L^2(U)} \|\nabla v^1\|_{L^2(U)} + A\|\nabla v^1\|_{L^2(U)} \|C_m\|_{L^2(U)}$$
$$+ \|C_m\|_{L^2(U)} \leq K\|C_m\|_{H^1_{per}(U)}.$$

Therefore
$$\|C'_m\|_{H^{-1}(U)} = \sup_{\|v\|=1}|<C'_m, v>|^2 \leq K\|C_m\|^2_{H^1_{per}(U)}.$$

Integrating again between 0 and T

$$\int_0^T \|C'_m\|^2_{H^{-1}(U)} dt \leq K\int_0^T \|C_m\|^2_{H^1_{per}(U)} dt \leq KT\|g\|^2_{L^2(U)}.$$

$\square$

4.1.4. Existence

Next we pass to limits as $m \to \infty$, to build a weak solution of our initial/boundary-value problem (4.2).

Theorem 4.1.5. *Under assumptions (4.1.3), there exists a weak solution of (4.2).*

Proof. According to the energy estimates (4.22), we see that the sequence $\{C_m\}_{m=1}^{\infty}$ is bounded in $L^2(0,T; H^1_{per}(U))$ and $\{C'_m\}_{m=1}^{\infty}$ is bounded in $L^2(0,T; H^{-1}(U))$. Consequently there exists a subsequence $\{C_{m_l}\}_{m_l=1}^{\infty} \subset \{C_m\}_{m=1}^{\infty}$ and a function $C \in H^1(0,T; H^1_{per}(U))$, with $C' \in L^2(0,T; H^{-1}(U))$, such that

$$\begin{aligned} C_{m_l} &\rightharpoonup C \quad \text{in} \in L^2(0,T; H^1_{per}(U)) \\ C'_{m_l} &\rightharpoonup C' \quad \text{in} \in L^2(0,T; H^1_{per}(U)). \end{aligned} \quad (4.23)$$

Next we fix an integer N and choose a function $v \in C^1([0,T]; H^1_{per}(U))$ having the form

$$v(t) = \sum_{k=1}^{N} d^k(t) w_k. \quad (4.24)$$

where $\{d^k\}_{k=1}^{N}$ are given smooth functions. We choose $m \geq N$, multiply (4.20) by $d^k(t)$, sum for $k = 1, \ldots, N$ and integrate with respect to t to find

$$\int_0^T [<C'_m, v> + (\ C_m,\ v)]\, dt = A \int_0^T (\ v, C_m\ W * C_m)\, dt + \int_{U_T} f(C_m)\, v. \quad (4.25)$$

We set $m = m_l$ and recall (4.23), to find upon passing to weak limits that

$$\int_0^T [<C', v> + (\ C,\ v)]\, dt = A \int_0^T (\ v, C\ W * C)\, dt + \int_{U_T} f(C)\, v. \quad (4.26)$$

This equality then holds for all functions $v \in L^2(0,T; H^1_{per}(U))$, as functions of the form (4.24) are dense in this space. Hence in particular

$$<C', v> + (\ C,\ v) = A(\ v, C\ W * C) + \int_U f(C)\, v \quad (4.27)$$

for each $v \in H^1_{per}(U)$ a.e. $0 \leq t \leq T$. Furthermore $u \in C([0,T]; L^2(U))$ (see [20]). In order to prove $u(0) = g$, we first note from (4.26) that

$$\int_0^T [-<v', C> + (\ C,\ v)]\, dt = \quad (4.28)$$
$$A \int_0^T (C\ W * C,\ v)\, dt + \int_{U_T} f(C)\, v + (C(0), v(0))$$

for each $v \in C^1([0,T] : H^1_{per}(U)$ with $v(T) = 0$. Similarly, from (4.25) we deduce

$$\int_0^T [-<v', C_m> + (\ C_m,\ v)]\, dt = \quad (4.29)$$
$$A \int_0^T (C_m\ W * C_m,\ v)\, dt + \int_{U_T} f(C_m)\, v + (C_m(0), v(0)).$$

Setting $m = m_l$ and employing again (4.23) we find

$$\int_0^T [-<v', C> + (\ C,\ v)]\, dt = A \int_0^T (C\ W * C,\ v)\, dt + \int_{U_T} f(C)\, v + (g, v(0)),$$

since $C_{m_l}(0) \to g$ in $L^2(U)$. As $v(0)$ is arbitrary, comparing (4.28) and (4.29), we conclude $C(0) = g$. □

4.2. Linear stability analysis

The formation of patterns stems from the fact that the uniform steady state can be unstable under particular conditions. A study of the parameters regime which leads to instability, can be useful to predict the behavior of the model, to have an estimate of some parameters and to understand their roles in details. In section ?? we already used the following modes ansatz which we prove in details here.

Theorem 4.2.1. *The functions*

$$u_{q_1}(x) = e^{iq_1 x}, \quad u_{q_2}(y) = e^{iq_2 y}, \quad z_n(\theta) = e^{in} \qquad (4.30)$$

where $q_1, q_2 \in \mathbb{R}$, if we are on a unbounded domain or $q_1, q_2 \in \mathbb{Z}$ otherwise and $n = 0, 1, 2, ...$, are the eigenfunctions of the operators present in the model 4.1.

Proof. The first two are eigenfunctions of $\frac{\partial^2}{\partial x^2} + \frac{\partial^2}{\partial y^2}$ with eigenvalues $\beta_q = -q^2$ ($q^2 = q_1^2 + q_2^2$), where as the third one of $\frac{\partial^2}{\partial \theta^2}$ with eigenvalues $\alpha_n = -n^2$. Indeed

$$\frac{\partial^2 e^{iqx}}{\partial x^2} = i^2 q^2 e^{iqx} = -q^2 e^{iqx}.$$

More difficult is to show that they are also eigenfunctions of the operator $W*$ with eigenvalues $\hat{W}(n, q_1, q_2) = \hat{W}_n \hat{W}_{q_1} \hat{W}_{q_2}$, where

$$\hat{W}_n = \int W_1(\theta) e^{-in\theta} d\theta, \quad \hat{W}_{q_j} = \int_0^L W_2(x) e^{-iq_j x} dx, \quad j = 1, 2 \qquad (4.31)$$

are respectively the Fourier coefficients of $W_1(\theta)$, $W_2(x)$ and $W_2(y)$. The separability of the function $\hat{W}(n, q_1, q_2)$ is a consequence of the separable nature of the kernel $W(\theta, \vec{x})$. Indeed

$$W*(e^{in\theta} e^{iq_1 x} e^{iq_2 y}) = (W_1(\theta) * e^{in\theta})(W_2(x) * e^{iq_1 x})(W_2(y) * e^{iq_2 y}). \qquad (4.32)$$

Starting with the first term, we want to show that $e^{in\theta}$ are the eigenfunctions and $\hat{W}_n$ the eigenvalues of the convolution operator, that means:

$$W_1 * \psi_n = \hat{W}_n \psi_n, \quad \text{for} \quad \psi_n(\theta) = e^{in\theta}. \qquad (4.33)$$

By the definition of convolution and with a change of variable we obtain

$$W_1 * \psi_n = \int W_1(\theta - \theta') e^{in\theta'} d\theta' = -\int_+^- W_1(\theta') e^{in(\theta-\theta')} d\theta' =$$
$$= \int_-^+ W_1(\theta') e^{in(\theta-\theta')} d\theta' = \int W_1(\theta') e^{in(\theta-\theta')} d\theta' \qquad (4.34)$$

Thus:

$$W_1 * \psi_n = e^{in\theta} \int W_1(\theta') e^{-in\theta'} d\theta' = \hat{W}_n \psi_n \qquad (4.35)$$

In the same way one can show that the eigenvalues of W_2* are the Fourier transform $\hat{W}_{q_1}$, $\hat{W}_{q_2}$ of $W_2(x)$, $W_2(y)$. $\square$

The eigenvalues of the exponential kernels can be easily calculated. The Fourier transform of the function $\gamma(x) = e^{-\frac{x^2}{2}}, \forall x \in \mathbb{R}$ is $e^{-\frac{q^2}{2}}, q \in \mathbb{R}$, whereas for $\gamma(ax) = e^{-\frac{(ax)^2}{2}}$ is $\frac{1}{a} e^{-\frac{q^2}{2a^2}}$ ([50]). As in our case $a = 1/\sigma$, we obtain

$$\hat{W}_{q_1} = \sqrt{\frac{2}{2\pi}} e^{-\frac{(q_1)^2}{2}}, \quad \hat{W}_{q_2} = \sqrt{\frac{2}{2\pi}} e^{-\frac{(q_2)^2}{2}}, \quad \hat{W}_n = \sqrt{\frac{1}{2\pi}} e^{-\frac{(n)^2}{2}}. \qquad (4.36)$$

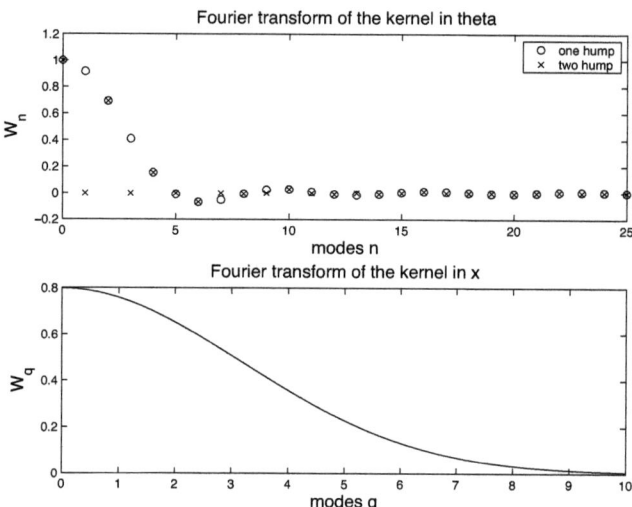

Figure 4.1.: *Eigenvalues of the convolution operators both in angle and in space. They both tend to zero when the number of modes increases.*

4.2.1. Linearization of the original model

Linearizing the model around the uniform steady state we get the following equation for the perturbation:

$$\frac{\partial C'_{n,q}}{\partial t} = \epsilon_1 \frac{\partial^2 C'_{n,q}}{\partial \theta^2} + \epsilon_2 \left(\frac{\partial^2 C'_{n,q}}{\partial x^2} + \frac{\partial^2 C'_{n,q}}{\partial y^2} \right) + \qquad (4.37)$$

$$- \frac{\partial}{\partial \theta} \left[\overline{C} \frac{\partial}{\partial \theta} \left(W * C'_{n,q} \right) + C'_{n,q} \frac{\partial}{\partial \theta} \left(W * \overline{C} \right) \right] +$$

$$- \frac{\partial}{\partial x} \left[\overline{C} \frac{\partial}{\partial x} \left(W * C'_{n,q} \right) + C'_{n,q} \frac{\partial}{\partial x} \left(W * \overline{C} \right) \right] +$$

$$- \frac{\partial}{\partial y} \left[\overline{C} \frac{\partial}{\partial y} \left(W * C'_{n,q} \right) + C'_{n,q} \frac{\partial}{\partial y} \left(W * \overline{C} \right) \right]$$

As $W * \overline{C} = \overline{C}$ for the normalization, it remains only

$$\frac{\partial C'_{n,q}}{\partial t} = \epsilon_1 \frac{\partial^2 C'_{n,q}}{\partial \theta^2} + \epsilon_2 \Delta_{xy} C'_{n,q} - \frac{\partial}{\partial \theta} \left[\overline{C} \frac{\partial}{\partial \theta} \left(W * C'_{n,q} \right) \right] \qquad (4.38)$$

$$- \frac{\partial}{\partial x} \left[\overline{C} \frac{\partial}{\partial x} \left(W * C'_{n,q} \right) \right] - \frac{\partial}{\partial y} \left[\overline{C} \frac{\partial}{\partial y} \left(W * C'_{n,q} \right) \right]$$

That gives:

$$\frac{\partial C'_{n,q}}{\partial t} = \epsilon_1 \frac{\partial^2 C'_{n,q}}{\partial \theta^2} + \epsilon_2 \Delta_{xy} C'_{n,q} - \overline{C} \frac{\partial^2}{\partial \theta^2} \left(W * C'_{n,q} \right) - \overline{C} \Delta_{xy} \left(W * C'_{n,q} \right) \qquad (4.39)$$

Then for each n and q we have:

$$\frac{\partial C'_{n,q}}{\partial t} = \left(-\epsilon_1 n^2 - \epsilon_2 q^2 + \overline{C} \hat{W}_q \hat{W}_n (n^2 + q^2) \right) C'_{n,q}. \qquad (4.40)$$

4.2.2. Analysis of the instability condition

Recalling that $\overline{C} = M$ is the total mass of the system, from the last equation we can get the instability condition:

$$\epsilon_2 q^2 < -\epsilon_1 n^2 + M \hat{W}_q \hat{W}_n (n^2 + q^2). \qquad (4.41)$$

Let the right hand side be $g_n(q) = -\epsilon_1 n^2 + M \hat{W}_q \hat{W}_n (n^2 + q^2)$. If we call $A = \frac{1}{2}M$ and write the expression (4.36) of $\hat{W}_q$, we obtain

$$g_n(q) = -\epsilon_1 n^2 + A \hat{W}_n (n^2 + q^2) e^{-\frac{(q)^2}{2}}.$$

We study this function separately in a bounded and unbounded spatial domain.

Unbounded spatial domain

Considering an infinite spatial domain, the wavenumber q is a continuous variable. As the Taylor expansion of $e^{-\frac{(q)^2}{2}}$ around zero is $1 - \frac{(q)^2}{2}$, neglecting the terms of order larger than two, we obtain the following asymptotes of the function g_n:

$$g_n(q) \approx \begin{cases} -\epsilon_1 n^2 + A \hat{W}_n q^2 + A \hat{W}_n n^2 \left(1 - \frac{(q)^2}{2} \right) & q \ll 1 \\ -\epsilon_1 n^2 + A \hat{W}_n q^2 e^{-\frac{(q)^2}{2}} & q \gg 1 \end{cases} \qquad (4.42)$$

Or better
$$g_n(q) \approx \begin{cases} (-\epsilon_1 + A\hat{W}_n)n^2 + A\hat{W}_n\left(1 - \frac{(n\sigma)^2}{2}\right)q^2 & q \ll 1 \\ -\epsilon_1 n^2 + A\hat{W}_n q^2 e^{-\frac{(q\sigma)^2}{2}} & q \gg 1 \end{cases} \quad (4.43)$$

This suggest us that the behavior of g_n depends on the sign of $(1 - \frac{(n\sigma)^2}{2})$. In fact, calculating the derivative of g_n, we obtain:

$$\begin{aligned} g'_n(q) &= A\hat{W}_n e^{-\frac{(q\sigma)^2}{2}}\left[q\sigma^2(n^2 + q^2) + 2q\right] \\ &= A\hat{W}_n e^{-\frac{(q\sigma)^2}{2}}\left[-q^3\sigma^2 + 2q\left(1 - \frac{(n\sigma)^2}{2}\right)\right], \end{aligned} \quad (4.44)$$

so that the function has a maximum in $q = 0$ if $(1 - \frac{(n\sigma)^2}{2}) < 0$ and $\hat{W}_n > 0$ (as in this case the second derivative is also negative), otherwise in

$$q = q^c = \sqrt{\frac{2}{\sigma^2} - n^2} = 0. \quad (4.45)$$

We can see in Fig. (4.1) that W_n is positive for some initial n and then it is null, giving stability to the model. If we use a double humped kernel, the odd modes are always stable. We can then distinguish three cases: $n = 0$ and for $n = 0$, $n^2 < \frac{2}{\sigma^2}$, $n^2 > \frac{2}{\sigma^2}$.

- $n = 0$:
  For large q we have stability, as $A\hat{W}_0 e^{-\frac{(q\sigma)^2}{2}}$ tends to zero. For small q there is instability if $\epsilon_2 < \epsilon_2^c = A\hat{W}_0$.

- $n^2 < \frac{2}{\sigma^2}$:
  The maximum of g_n occurs for $q = q^c = 0$. For large q we have stability. For small q there is stability if $\epsilon_i > \epsilon_i^c$. The critical value of ϵ can be calculated from the first asymptote:

$$\epsilon_1^c = A\hat{W}_n \quad \epsilon_2^c = A\hat{W}_n\left(1 - \frac{(n\sigma)^2}{2}\right). \quad (4.46)$$

In fact the instability condition is given by;

$$\left[\epsilon_2^c - A\hat{W}_n\left(1 - \frac{(n\sigma)^2}{2}\right)\right]q^2 < \left(-\epsilon_1 + A\hat{W}_n\right)n^2. \quad (4.47)$$

For ϵ_2 large and ϵ_1 small we can have instability for small q. On the other hand, if $\epsilon_2 < \epsilon_2^c$ and $\epsilon_1 > \epsilon_1^c$, instability can occur for $q = q^c$. There is still stability if ϵ_2 is large enough and ϵ_1 not too big, but instability if ϵ_2 is small enough. Summarizing we get

  – $\epsilon_i > \epsilon_i^c$: stability,
  – $\epsilon_1 < \epsilon_1^c$, $\epsilon_2 > \epsilon_2^c$: instability at small q,
  – $\epsilon_1^c < \epsilon_1 < \epsilon_1'$, $\epsilon_2 < \epsilon_2^c$: instability at $q = q^c$ if $\epsilon_2 < \epsilon_2'(\epsilon_1)$. To calculate ϵ_1', ϵ_2' a transcendental equation has to be solved.

- $n^2 > \frac{2}{\sigma^2}$:
  In this case, the maximum of g_n is at $q = 0$, then the function decreases. We observe that ϵ_2^c is negative. For ϵ_1 small enough ($\epsilon_1 < \epsilon_1^c$), g_n remains below $\epsilon_2 q^2$ for all q. Independently from ϵ_2 we have

  – stability for large ϵ_1 ($> \epsilon_1^c$)
  – instability for small ϵ_1 ($< \epsilon_1^c$) and small q.

Scenarios

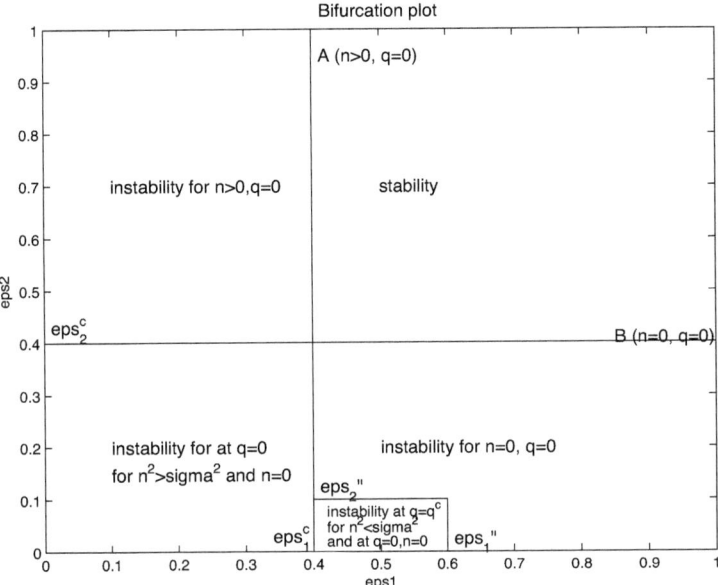

Figure 4.2.: *Bifurcation diagram: A and B are the line of primary bifurcations corresponding respectively to the mode $n > 0$, $q = 0$ and $n = 0$, $q = 0$.*

Then these three scenarios may occur:

- We first cross the line A: the mode $n > 0, q = 0$ breaks the stability. Angular patterns evolve whereas homogeneity in space remains. The cells then orient themselves, but without aggregating.
- We first cross the line B: the mode $n = 0, q = 0$ breaks the stability. The angular disordered pattern with the spatial inhomogeneities evolves. This means that the objects aggregate but they do not align.
- Scenario C: The mode $n > 0, q = q^c$ breaks the stability. In this case the spatial density is not altered, but the angle of preferred orientation changes periodically in space creating squares or hexagons.

Bounded domain

In this case our variable q is a natural number, $q = 0, 1, \ldots$. We can distinguish these cases:

- $q = 0$: The instability condition does not depend on ϵ_2 and is given by:

$$\epsilon_1 < A\hat{W}_n \tag{4.48}$$

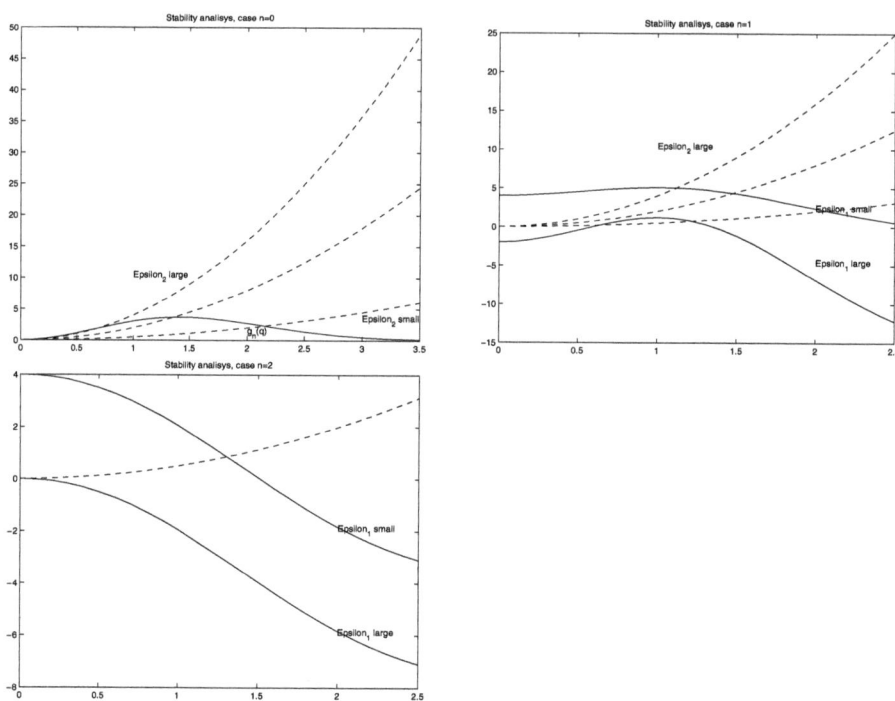

Figure 4.3.: *Stability plots for unbounded one-dimensional domain and $^2 = 1$. The dashed lines and the solid lines show the left-hand side and the right-hand side of the instability condition, respectively. We have instability when the dashed lines are below the solid ones.*

- $n = 0$: The instability condition does not depend on ϵ_1 and is given by:

$$\epsilon_2 < A\hat{W}_n \qquad (4.49)$$

- $n, q > 0$: This case is very similar to the continuum one.

We do not observe large dierences in the two cases of bounded and unbounded domain.

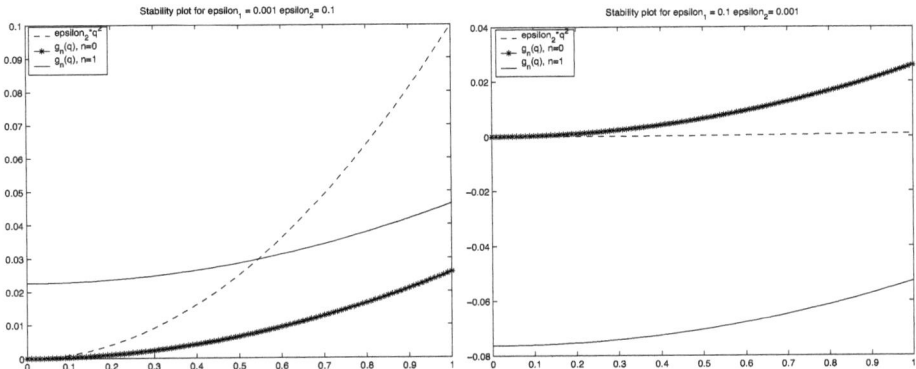

Figure 4.4.: *Stability plots for the model with $M = 1$, $^2 = \frac{1}{50}$. In the first case is the mode $n = 1$ that breaks the stability, in the second case the mode $n = 0$.*

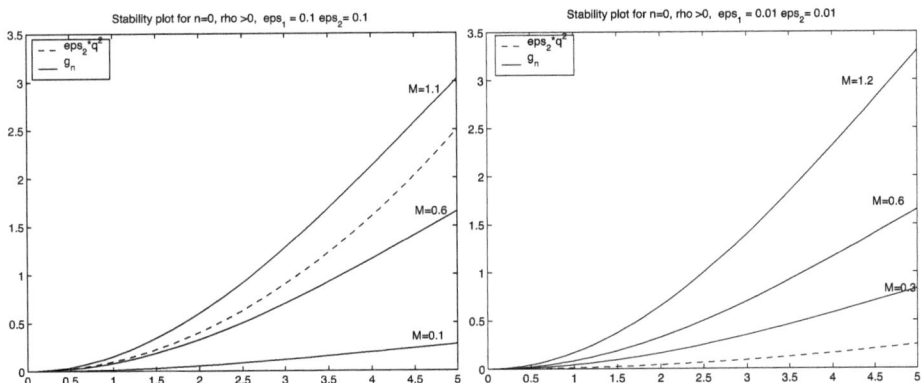

Figure 4.5.: *Stability plots for the model with big and small epsilons. The dashed lines and the solid lines show the left-hand side and the right-hand side of the instability condition, respectively. We have instability when the dashed lines are below the solid ones. In the first case we have instability from the beginning, in the second one after the mass M reaches the value of one.*

4.2.3. Linearization of the extended model

In this case the equation for the perturbation is

$$\frac{\partial C'_{n,q}}{\partial t} = \epsilon_1 \frac{\partial^2 C'_{n,q}}{\partial \theta^2} + \epsilon_2 \left(\frac{\partial^2 C'_{n,q}}{\partial x^2} + \frac{\partial^2 C'_{n,q}}{\partial y^2} \right) + \rho \left(1 - 2\frac{\overline{C}}{K} \right) C'_{n,q}$$
$$- \gamma_2 \overline{C} \left\{ \frac{\partial}{\partial x} \left[\frac{\partial}{\partial x} \left(W * C'_{n,q} \right) \right] + \frac{\partial}{\partial y} \left[\frac{\partial}{\partial y} \left(W * C'_{n,q} \right) \right] \right\}$$
$$- \gamma_1 \overline{C} \frac{\partial}{\partial \theta} \left[\frac{\partial}{\partial \theta} \left(W * C'_{n,q} \right) \right]. \quad (4.50)$$

This time we also need the equation for the uniform steady state, which depends on time:

$$\frac{\partial \overline{C}}{\partial t} = \rho \overline{C} \left(1 - \frac{\overline{C}}{K} \right). \quad (4.51)$$

With some calculations (see chapter 3) for large times we get the instability condition

$$\epsilon_2 q^2 < -\epsilon_1 n^2 + K \hat{W}_n \hat{W}_q (\gamma_1 n^2 + \gamma_2 q^2) - \rho \quad (4.52)$$

If we take $\gamma_1 = \gamma_2 = 1$ and compare (4.52) to the condition analyzed in the previous chapter, we see that a factor ρ is now present and that $\overline{C}$ is replaced by K. Recalling that $\hat{W}_q = \frac{1}{2} e^{-\frac{(q)^2}{2}}$, if we let $A = \frac{1}{2} K$, we can distinguish the two asymptotes

$$g_n(q) \approx \begin{cases} (-\epsilon_1 + A \hat{W}_n) n^2 + A \hat{W}_n \left(1 - \frac{(n)^2}{2} \right) q^2 - \rho & q << 1 \\ -\epsilon_1 n^2 + A \hat{W}_n q^2 e^{-\frac{(q)^2}{2}} - \rho & q >> 1. \end{cases} \quad (4.53)$$

As the derivative of g_n is independent of ρ, we can distinguish for $\rho > 0$ the same three cases as before:

- $n = 0$:
  For small q, we have instability for all ϵ_1, if

$$\epsilon_2 < \epsilon_2^c = A \hat{W}_0 e^{-\frac{(q)^2}{2}} - \frac{\rho}{q^2}. \quad (4.54)$$

  For large q there is stability for $\epsilon_2 > A\hat{W}_0$. For q small enough, $\rho/q^2 > A\hat{W}_0$, so that ϵ_2 will always be larger then $A\hat{W}_0 - \rho/q^2$.

- $n^2 < \frac{2}{2}$: see previous section.

- $n^2 > \frac{2}{2}$:
  We have only a maximum in $q = 0$. For $\epsilon_1 > \epsilon_1^c$ we have stability, otherwise we can have instability if $(\epsilon_1^c - \epsilon_1) n^2 > \rho$.

Integral in the logistic term

If we consider the model with the integral in the logistic term

$$\rho C \left(1 - \frac{\int_0^2 C(t, x, \theta) d\theta}{K} (L_x L_y) \right), \quad (4.55)$$

the stability condition changes slightly. The term $O(1)$ remain the same (4.51), but the logistic term $O(\epsilon)$ is now:

$$\rho\left(1 - \frac{\overline{C}L_xL_y 2\pi}{K}\right)C'_{n,q} - \rho\frac{\overline{C}\int_0^2 C'_{n,q}d\theta}{K}L_xL_y. \qquad (4.56)$$

But for the choice of $C'_{n,q}$, its integral is null, so that remains the term

$$\rho\left(1 - \frac{\overline{C}}{K}L_xL_y 2\pi\right)C'_{n,q} \qquad (4.57)$$

which has a dierence of factor $2\pi L_x L_y$ from the previous, leading to the instability condition:

$$\epsilon_2 q^2 < -\epsilon_1 n^2 + \frac{K}{2\pi L_x L_y}\hat{W}_n\hat{W}_q(\gamma_1 n^2 + \gamma_2 q^2) - \rho \qquad (4.58)$$

Analysis of the parameters

Because in our experiments we observed the formation of arrays without the presence of aggregation we are interested in patterns with $n > 0$ and q small or null. Let for simplicity $K = \frac{K}{2\pi L_x L_y}$. As we saw, for $n > 0$ there are two cases: $n^2 < 2/\sigma^2$ and $n^2 > 2/\sigma^2$. Let's here for simplicity consider a kernel so that $W_n = 0$ for $n > 4$. Then the only unstable modes can be $n = 1, 2, 3, 4$. For all n there is a critical value σ^c so that

- if $\sigma^2 < \sigma^c$ the unstable mode in space is $q = q^c = \sqrt{\frac{2}{2} - n^2}$ for $\epsilon_1 < \epsilon_1^c = A\hat{W}_n$ and $\epsilon_2 > \epsilon_2^c = A\hat{W}_n\left(1 - \frac{(n)^2}{2}\right)$;

- if $\sigma^2 > \sigma^c$ the unstable mode in space is $q = 0$ for $\epsilon_1 < \epsilon_1^c = A\hat{W}_n$ and all ϵ_2.

The critical value σ^c depends on n:

n	1	2	3	4
σ^c	$+\infty$	0.5	0.2	0.1

We note that with a two humped kernel the only angular modes that can be unstable are $n = 2$ and $n = 4$.

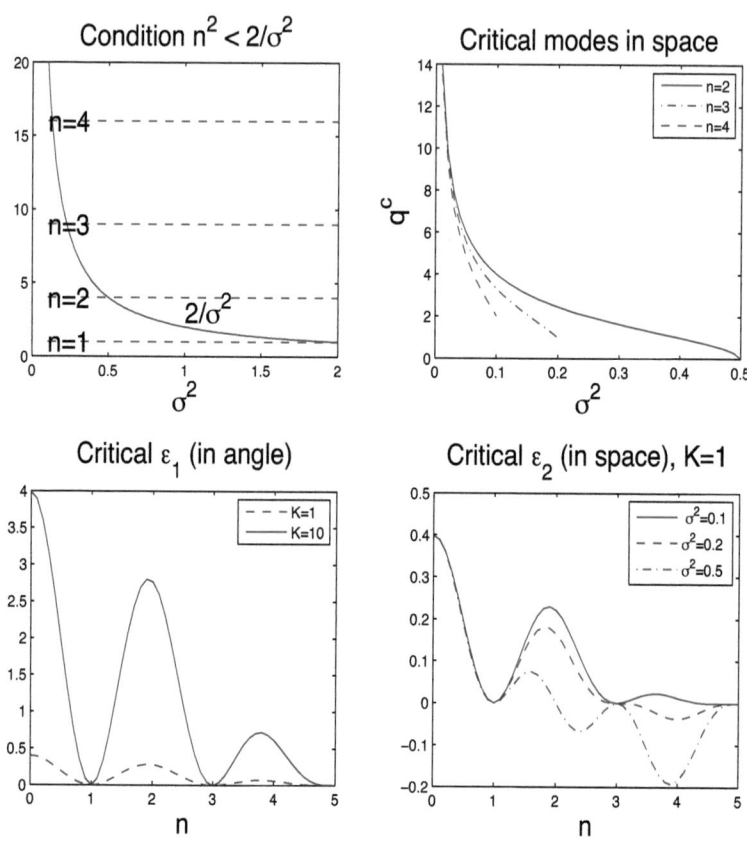

Figure 4.6.: *Critical values. In the first frame we plot the function $\frac{2}{\sigma^2}$, in the second one the value of the critical mode in space (4.45). In the second row we plot the critical diffusion coefficients for different values of K (4.46).*

4.3. Numerical methods

4.3.1. Integration in space and angle

The domain we consider is very small compared with the whole well, so that the boundary of the well should not influence our domain. It is then reasonable to consider periodic boundary conditions both in space and in angle. The problem in consideration is not solvable analytically, but numerically. Our goal is to reduce it to a discrete problem that we are able to solve. We begin by discretizing the spatial and angular domains. For convenience, we will use a uniform grid. Let set $h_x = L_x/(N-1)$, $x_j = j h_x$, $j = 0, ..., N-1$; $h_y = L_y/(M-1)$, $y_j = j h_y$ $j = 0, ..., M-1$ and $h = 2\pi/(Q-1)$, $\theta_j = j h - \pi$, $j = 0, ..., Q-1$, $N, M, Q \in \mathbb{Z}$. At each time t fixed, our problem is to solve a differential equation in space. For the nonlinearity of the system, the simplest method to use is the centered finite difference. As all the terms can be treated at the same way, let's consider only the variable x and call h the step and L the length of the domain. Applying the centered difference method [46] to our equation, we obtain that the second derivative of C can be approximated by

$$\frac{\partial^2 C}{\partial x^2}(x_j) \approx \frac{C_{j+1} - 2C_j + C_{j-1}}{h^2} + O(h^2). \tag{4.59}$$

where $C_j = C(x_j, y, \theta, t)$, for fix y, θ and t. Pretending that there are two additional points $x_{-1} = -h$ and $x_N = L + h$ and imposing the periodic conditions $C_N = C_0$, $C_{-1} = C_{N-1}$ for each θ, y and t, the second derivative can be calculated as a product of a vector and a matrix. The derivatives of the kernels W_1 and W_2 can be easily calculated analytically. Moreover, if we write $F(C) = C(K * C)$, where K is the kernel, the first derivative of F can be computed with centered finite difference too:

$$\frac{\partial F}{\partial x}(x_j) = \frac{F_{j+1} - F_{j-1}}{2h} + O(h^2), \tag{4.60}$$

again with periodic boundary conditions.

The convolution is computed with the trapezoidal rule which for a function f defined on the grid is given by

$$Q(f) := \int_0^L f(x) \approx Q_N(f) = h \left[\frac{1}{2}(f(x_0) + f(x_{N-1})) + \sum_{j=1}^{N-2} f(x_j) \right]. \tag{4.61}$$

In the next sections we show that with the composite trapezoidal rule the integration error decays at least exponentially if we integrate periodic analytic functions [30].

Theorem 4.3.1. Let $f \in C^2[0, L]$. Then the remainder

$$R_T(f) := \int_0^L f(x)dx - h \left[\frac{1}{2}(f(x_0) + f(x_{N-1})) + \sum_{j=1}^{N-2} f(x_j) \right] \tag{4.62}$$

for the composite trapezoidal rule can be estimated by

$$|R_T(f)| \leq \frac{1}{12} h^2 L \|f''\|_\infty. \tag{4.63}$$

Proof. Define the Peano kernel

$$K_T(x) := \frac{1}{2}(x - x_{j-1})(x_j - x), \quad x_{j-1} \leq x \leq x_j, \tag{4.64}$$

for $j = 0, \ldots, N-1$. Then, straightforward partial integrations yield

$$\int_0^L K_T(x) f''(x) dx = -R_T(f). \tag{4.65}$$

Now the estimate follows from

$$\int_0^L K_T(x) dx = \frac{h^2}{12} L \tag{4.66}$$

and the observation that K_T is nonnegative on $[0, L]$. □

If the function is smooth and periodic, we can obtain better results. To proof this we have to introduce the trigonometric interpolation. Let call $C(G)$ the space of continuous real- or complex-valued functions on compact subsets $G \subset \mathbb{R}^m$.

Theorem 4.3.2. *Let $U_N \subset C(G)$ be an N-dimensional subspace and let $x_0, \ldots, x_{N-1}$ be n points in G such that U_N for each function from U_N that vanishes in these points, vanishes identically. Then, given n values $f_0, \ldots, f_{N-1}$, there exists a uniquely determined function $u \in U_N$ with the interpolation property $u(x_j) = f_j$, for $j = 0, \ldots, N-1$. We call $P_N : C(G) \to U_N$ the interpolation operator for the interpolation data $f_j = f(x_j)$, for $j = 0, \ldots, N$ that is defined by the mapping $f \to u$.*

Proof. Let $U_n = \text{span}\{u_0, \ldots, u_{N-1}\}$. Then the solution to the interpolation problem is given by $u = \sum_{k=0}^{N-1} \gamma_k u_k$, where the coefficients $\gamma_0, \ldots, \gamma_{N-1}$ are determined by the uniquely solvable linear system $\sum_{k=0}^{N-1} \gamma_k u_k(x_j) = f_j$, for $j = 0, \ldots, N-1$. Let $L_0, \ldots, L_{N-1}$ denote the Lagrange basis for U_N, i.e., we have the interpolation property $L_k(x_j) = \delta_{jk}$, for $j, k = 0, \ldots, N-1$, where $\delta_{jk} = 1$ for $k = j$, and $\delta_{jk} = 0$ for $k \neq j$. Then

$$P_N g = \sum_{k=0}^{N-1} f(x_k) L_k. \tag{4.67}$$

□

In particular, let $t_j = \frac{j}{N}, j = 0, \ldots, 2N-1$ be an equidistant subdivision of the interval $[0, 2\pi]$ with an even number of grid points. Then, given the values $f_0, \ldots, f_{2N-1}$, there exists a unique trigonometric polynomial of the form

$$u(t) = \frac{\alpha_0}{2} + \sum_{k=1}^{N-1} [\alpha_k \cos kt + \beta_k \sin kt] + \frac{\alpha_N}{2} \cos N t \tag{4.68}$$

with the interpolation property $u(t_j) = f_j, j = 0, \ldots, 2N-1$. Its coefficients are given by

$$\alpha_k = \frac{1}{N} \sum_{j=0}^{2N-1} f_j \cos kt_j, \quad k = 0, \ldots, N,$$

$$\beta_k = \frac{1}{N} \sum_{j=0}^{2N-1} f_j \sin kt_j, \quad k = 1, \ldots, N-1.$$

From this we deduce that the Lagrange basis of the trigonometric interpolation has the form

$$L_j(t) = \frac{1}{2N} \left\{ 1 + 2 \sum_{k=1}^{N-1} \cos k(t - t_j) + \cos N(t - t_j) \right\}, \tag{4.69}$$

for $t \in [0, 2\pi]$ and $j = 0, \ldots, 2N - 1$. Using the real part of the geometric sum

$$1 + 2\sum_{k=1}^{m-1} e^{ikt} + e^{imt} = i(1 - e^{imt})\cot\frac{t}{2}, \quad 0 < t < 2\pi \tag{4.70}$$

we can transform (4.69) into

$$L_j(t) = \frac{1}{2N}\sin N(t - t_j)\cot\frac{t - t_j}{2}, \quad t = t_j. \tag{4.71}$$

Theorem 4.3.3. *Let* $f : \mathbb{R} \to \mathbb{R}$ *be analytic and* $2\pi-$ *periodic. Then there exists a strip* $D = \mathbb{R} \times (-s, s) \subset \mathbb{C}$ *with* $s > 0$ *such that* f *can be extended to a holomorphic and* $2\pi-$ *periodic bounded function* $f : D \to \mathbb{C}$. *The error*

$$R_T(f) := \frac{1}{2\pi}\int_0^{2\pi} f(t)\,dt - \frac{1}{2N}\sum_{j=0}^{2N-1} f\left(\frac{j\pi}{N}\right) \tag{4.72}$$

for the composite trapezoidal rule can be estimated by

$$|R_T(f)| \leq M(\coth(Ns) - 1), \tag{4.73}$$

where M *denotes a bound for the holomorphic function* f *on* D.

Proof. Because $f : \mathbb{R} \to \mathbb{R}$ is analytic, at each point $t \in \mathbb{R}$ the Taylor expansion provides a holomorphic extension of f into some open disk in the complex plane with radius $r(t) > 0$ and center t. The extended function again has period 2π, since the coefficients of the Taylor series at t and $t + 2\pi$ coincide for the $2\pi-$ periodic function $f : \mathbb{R} \to \mathbb{R}$. The disks corresponding to all points of the interval $[0, 2\pi]$ provide an open covering of $[0, 2\pi]$. Being this interval compact, a finite number of these disks suffices to cover it. Then we have an extension into a strip D with finite width $2s$ contained in the union of the finite number of disks. Without loss of generality we may assume that f is bounded on D.

Let $0 < \sigma < s$ be arbitrary and $= [-\frac{\pi}{2N}, 2\pi - \frac{\pi}{2N}] \times [-\sigma, \sigma]$. We want to show that

$$\frac{1}{2\pi i}\int \frac{\cot\frac{t-\tau}{2}}{\sin N\tau}f(\tau)d\tau = \frac{2f(t)}{\sin Nt} - \frac{1}{N}\sum_{j=0}^{2N-1}(-1)^j f(t_j)\cot\frac{t - t_j}{2} \tag{4.74}$$

for $-\pi/2N \leq t \leq 2\pi - \pi/2N$, $t = t_k$, $k = 0, \ldots, 2N - 1$.

We recall that the integral of $g(\tau) = \frac{\cot\frac{t-\tau}{2}}{\sin N\tau}f(\tau)$ along this curve, can be calculated with the residue theorem

$$\int g(\tau)d\tau = 2\pi i\sum_{j=1}^n I(\,,j)\mathrm{Res}(g, \tau_j), \tag{4.75}$$

being τ_j the singular points of g and $I(\,,j)$ the winding number of the curve. If we consider it counterclockwise oriented, $I(\,,j) = -1$ for all j. For functions of the form $g(\tau) = q(\tau)/p(\tau)$, where $q(\tau)$ is regular and $p(\tau)$ has a simple zero in τ_0, the residue can be calculated as:

$$\mathrm{Res}(g(\tau_0)) = \frac{q(\tau_0)}{p'(\tau_0)}. \tag{4.76}$$

We observe that inside , $\sin(N\tau)$ has the simple zeros $t_j = j\pi/N$, for $j = 0, \ldots, 2N - 1$, and $\sin(\frac{t-\tau}{2})$ the simple zero $\tau = t$. Applying formula (4.76) we can calculate our residues

$$\mathrm{Res}\left(\cot\frac{t-\tau}{2}, t\right) = \mathrm{Res}\left(\frac{\cos\frac{t-\tau}{2}}{\sin(\frac{t-\tau}{2})}, t\right) = \frac{\cos 0}{-\frac{1}{2}\cos 0} = -2,$$

$$\mathrm{Res}\left(\frac{1}{\sin(N\tau)}, t_j\right) = \frac{1}{N\cos j\pi} = \frac{1}{N}(-1)^j, \quad j = 0, \ldots, 2N - 1$$

and then calculate the integral (4.74) substituting these values in (4.75). Hence, in view of (4.71) and observing that $\sin N(t-t_j) = \sin Nt - \pi j$, and then $\sin N(t-t_j)/\sin Nt = (-1)^j$, we obtain that the remainder term of the trigonometric interpolation is

$$f(t) - (P_N f)(t) = \frac{\sin Nt}{4\pi i} \int \frac{\cot\frac{t-\tau}{2}}{\sin N\tau} f(\tau) d\tau \qquad (4.77)$$

where we can obviously drop the restriction that t does not coincide with an interpolation point. From the periodicity of the integrand and since, by the Schwarz reflection principle, f enjoys the symmetry property $f(\bar{\tau}) = \overline{f(\tau)}$ and $\cot(-x) = -\cot(x)$, we find the representation

$$f(t) - (P_N f)(t) = \frac{1}{2\pi} \sin Nt \, \text{Re} \left\{ \int_i^{i+2} \frac{i\cot\frac{-t}{2}}{\sin N\tau} f(\tau) d\tau \right\}. \qquad (4.78)$$

Now, integrating the geometric series (4.70) we obtain

$$\int_0^2 e^{imt} \cot\frac{t}{2} dt = 2\pi i.$$

Since $R_T(f) = \int_0^2 (f(t) - (P_N f)(t)) dt$, integrating (4.78) and using this last expression, we have

$$R_T(f) = \frac{1}{2\pi} \text{Re} \left\{ \int_i^{i+2} (1 - i\cot(N\tau)) f(\tau) d\tau \right\} \qquad (4.79)$$

for all $0 < \sigma < s$. Finally, the estimate follows from $|1 - i\cot(N\tau)| \leq \cot(N\sigma) - 1$ for $\text{Im}\,\tau = \sigma$ and passing to the limit $\sigma \to s$. $\square$

We can summarize this theorem by the estimate

$$|R_T(f)| \leq Ce^{-2Ns} \qquad (4.80)$$

for the composite trapezoidal rule for periodic analytic functions, where C and s are some positive constants depending on f, i.e., the integration error decays at least exponentially.

Applying the quadrature rule to the convolution term, we see that it can again be written as a product of a matrix and a vector

$$h_x \begin{pmatrix} \frac{1}{2}V(x_0-x_0) & \frac{1}{2}V(x_1-x_0) & \cdots & \frac{1}{2}V(x_{N-1}-x_0) \\ V(x_0-x_1) & V(x_1-x_1) & \cdots & V(x_{N-1}-x_1) \\ \vdots & \vdots & \vdots & \vdots \\ \frac{1}{2}V(x_0-x_{N-1}) & \frac{1}{2}V(x_1-x_{N-1}) & \cdots & V(x_{N-1}-x_{N-1}) \end{pmatrix} \begin{pmatrix} C_0 \\ C_1 \\ \vdots \\ C_{N-2} \\ C_{N-1} \end{pmatrix}.$$

4.3.2. Integration in time

Let write our system for each (x, y, θ) fixed, as

$$\frac{\partial C}{\partial t} = f(C), \quad C(x, y, \theta, 0) = C_0(x, y, \theta), \qquad (4.81)$$

in order to use methods which solve the equation

$$\frac{\partial y}{\partial t} = f(y), \quad y(0) = y_0. \qquad (4.82)$$

Applying explicit methods one should be aware that restrictions on the time step are needed. The region that contain the value for which there is stability is called stability region and is defined as $S = \{z \in \mathbb{C} : |R(z)| \leq 1\}$, where $R(z)$ is the stability function, that can be interpreted as the numerical solution after one step of the Dahlquist test equation

$$y' = \lambda y, \quad y_0 = 1, \quad z = \Delta t \lambda. \tag{4.83}$$

For a nonlinear system, λ is the largest eigenvalue of the matrix obtained linearizing the system. To have stability $z = \Delta t \lambda$ should stay in S. If λ is real (otherwise we take its real part), this means that $\Delta t \lambda < l$, where l is the largest intersection of S with the real axis, and of course depends on the method. Considering the diffusion terms approximated by a finite difference method, the stability of a numerical method depends on the eigenvalues of the correspondent matrix. From the theorem of Gershgorin [12],

Theorem 4.3.4 (Gershgorin). *Let* $A = \{a_{ij}\}$ *be a square matrix of order* n. *For any* $k = 1, 2, \ldots, n$, *consider* $D_k \subset \mathbb{C}$ *defined by*

$$|z - a_{kk}| \leq \quad k = \sum_{\substack{j=1 \\ j \neq k}}^{n} |a_{kj}|. \tag{4.84}$$

The eigenvalues of A *are in the union of the* n *disk* D_k.

Proof. Let λ be an eigenvalue and u an associated eigenvector of A, so that

$$\max_{1 \leq n} |u_i| = |u_k| = 1.$$

The kth equation of the system $Au = \lambda u$ is

$$(\lambda - a_{kk})u_k = \sum_{\substack{j=1 \\ j \neq k}}^{n} a_{kj} u_j.$$

Since $|u_j| \leq |u_k| = 1$, if we take the modules we obtain

$$|\lambda - a_{kk}| = |\sum_{\substack{j=1 \\ j \neq k}}^{n} a_{kj} u_j| \leq \sum_{\substack{j=1 \\ j \neq k}}^{n} |a_{kj}||u_j| \leq \sum_{\substack{j=1 \\ j \neq k}}^{n} |a_{kj}|.$$

Then $\lambda \in D_k$. □

From this theorem we know that the largest eigenvalue of a square matrix cannot exceed the largest sum of the elements (taken with their modules) along any row or any columns ([44]). In our matrix then the largest eigenvalue is smaller then $4/h_x^2$. Considering also the diffusion coefficients and all three diffusion terms, the time step has to satisfy

$$4\Delta t \left(\frac{\epsilon_1}{h^2} + \frac{\epsilon_2}{h_x^2} + \frac{\epsilon_3}{h_y^2} \right) < l. \tag{4.85}$$

In conclusion we should take a time step

$$\Delta t < \frac{l h^2 h_x^2 h_y^2}{4(\epsilon_1 h_x^2 h_y^2 + \epsilon_2 h^2 h_y^2 + \epsilon_3 h_x^2 h^2)} \tag{4.86}$$

Theorem 4.3.5. *The diusion terms are the only terms that play a role in the st ability of a numerical method applied on equation (4.1).*

Proof. Let consider the linearized equation in one dimension

$$\frac{\partial C}{\partial t} = \epsilon \frac{\partial^2 C}{\partial x^2} - \overline{C}\frac{\partial^2}{\partial x^2}(K*C). \tag{4.87}$$

Applying a finite dierence method and calculating the convo lution with the trapezoidal rule, the second term can be approximated by the sum of three terms, each be the product of a matrix and the vector $(C_0, C_1, \ldots, C_{N-1})^T$.

$$\frac{\partial^2}{\partial x^2}(V*u)(x_i) = \frac{\overline{C}}{h_x^2}\left((V*u)(x_{i-1}) - 2(V*u)(x_i) + (V*u)(x_{i+1})\right). \tag{4.88}$$

One matrix is given by the matrix 4.3.1 multiplied by $-2\overline{C}/h_x$, and the other two are given by the matrix 4.3.1 with a shift in the rows multiplied by $\overline{C}/h_x$. The fact that the matrixes are multiplied by a factor $1/h_x$, whereas for the diusion we have a factor of $1/h_x^2$, suggests us already that this terms don't play a decisive role. The matrixes have all real eigenvalues and for the normalization of the kernels and the first Gerschgorin's theorem the largest eigenvalue of the linearized equation is given by

$$\lambda \leq 4\epsilon h_x^2 + \frac{4\overline{C}}{h_x} \tag{4.89}$$

From this we can calculate the restriction on Δt

$$\Delta t \leq \frac{l}{4}h_x^2\left(\frac{1}{\epsilon + \overline{C}h_x}\right). \tag{4.90}$$

The constant $\overline{C}$ is given by the total mass of the system, so that for h_x su cient small $\epsilon + \overline{C}h_x \approx \epsilon$. □

Runge-Kutta method

Let Δt be our time step, b_i, a_{ij} real numbers and $c_i = \sum_{j=1}^{s}$. An explicit Runge-Kutta method of stage s is given by

$$\begin{aligned} g_0 &:= y_0, \quad g_i = y_0 + \Delta t \sum_{j=1}^{i-1} a_{ij}k_j \quad i = 1,\ldots,s \\ k_i &= f(t_0 + c_i\Delta t, g_i) \quad i = 1,\ldots,s \\ y_1 &:= y_0 + \Delta t \sum_{i=1}^{s} b_i k_i \end{aligned} \tag{4.91}$$

An example is the fourth order method

$$\begin{aligned} k_1 &= f(y_0); \\ k_2 &= f(y_0 + (\Delta t/2)k_1); \\ k_3 &= f(y_0 + (\Delta t/2)k_2); \\ k_4 &= f(y_0 + \Delta t k_3); \\ y_1 &= y_0 + (\Delta t/6)(k_1 + 2k_2 + 2k_3 + k_4); \end{aligned} \tag{4.92}$$

For this method, we need the time step Δt to be su cient small in order to gain stability.

Theorem 4.3.6. *For a Runge-Kutta method of order p the stability function is given by*

$$R(z) = 1 + z + \frac{z^2}{2!} + \dots + \frac{z^p}{p!}, \tag{4.93}$$

Proof. The exact solution of (4.83) is e^z and therefore the numerical solution $y_1 = R(z)$ must satisfy

$$e^z - R(z) = O(\Delta t^{p+1}) = O(z^{p+1}). \tag{4.94}$$

□

If we want take a time step larger, we have to seek for methods with a larger stability region. Within the explicit RK family the design methods are rather specials. Instead of constructing methods with a maximal order of consistency for a minimal number of stage, for stabilized methods the greatest interest consist in having a maximal stability interval and a low order.

Chebyshev methods

In particular we would like a method with a larger l ($-\Delta t \lambda > -l$).

Theorem 4.3.7. *The optimal stability function for a first order method, with stage s, is given by*

$$R_s(z) = T_s\left(1 + \frac{z}{s^2}\right), \tag{4.95}$$

where $T_s(x)$ are the Chebyshev polynomials defined as

$$T_0(x) = 1, \quad T_1(x) = x, \quad T_s(x) = 2xT_{s-1}(x) - T_{s-2}(x) \quad s > 1. \tag{4.96}$$

Moreover, with this stability function the maximal l is $l = 2s^2$.

Proof. $R_s(x)$ is an optimal stability function if $|R_s(x)| \leq 1$. Now, we first observe that the Chebyshev polynomial can also be written as $T_s(x) = \cos(s \arccos x)$ for $x \in [-1, 1]$. Indeed, for $x = \cos\theta$ we have $T_s(\cos\theta) = \cos s\theta$. Then

$$\begin{aligned}
T_0(x) &= T_0(\cos\theta) = \cos 0\theta = 1 \\
T_1(x) &= T_1(\cos\theta) = \cos\theta \\
T_2(x) &= T_2(\cos\theta) = \cos 2\theta = 2\cos\theta\cos\theta - 1 = 2xT_1(x) - T_0(x) \\
T_3(x) &= T_3(\cos\theta) = 2\cos\theta\cos 2\theta - \cos\theta = 2xT_2(x) - T_1(x),
\end{aligned}$$

and so on we obtain exactly (4.96). Then

$$\left|T_s\left(1 + \frac{z}{s^2}\right)\right| \leq 1 \iff \left|1 + \frac{z}{s^2}\right| \leq 1 \iff z \in [-2s^2, 0].$$

It remains to show that $l = 2s^2$ is optimal. The grad of $R_s(z)$ is s and $R_s(0) = T_s(1) = 1$, $R'_s(0) = T'_s(1)\frac{1}{s^2} = 1$. Then

$$R_s(x) = 1 + x + \alpha_2 x + \dots \alpha_s x^s$$

for some coefficients α_i. Now, there exist $s + 1$ points $y_k = \cos(\frac{k}{s})$ where $T_s(y_k) = (-1)^k$, for $k = 0, \dots s$. Then if it exists a $\tilde{l} \geq l$ and a corresponding $\tilde{R}_s(z) = 1 + z + \tilde{\alpha}_2 z^2 + \dots + \tilde{\alpha}_s z^s$ with $|\tilde{R}_s(z)| \leq 1$, $\tilde{R}_s(z) - R_s(z)$ should have at least $s - 1$ zeros in $(-\tilde{l}, 0)$. But $\tilde{R}_s(z) - R_s(z) = z^2(\tilde{\alpha}_2 - \alpha_2 + (\tilde{\alpha}_3 - \alpha_3)z + \dots + (\tilde{\alpha}_{s-2} - \alpha_{s-2})z^{s-2} = z^2 p_{s-2}(z)$. But the polynomial $p_{s-2}(z)$ cannot have $s - 1$ zeros.

□

Theorem 4.3.8. *A Runge-Kutta method with such a stability function is given by*

$$g_0 := y_0, \quad g_1 := y_0 + (1/s^2)\Delta t f(g_0),$$
$$g_i := (2/s^2)\Delta t f(g_{i-1}) + 2g_{i-1} - g_{i-2}, \quad (4.97)$$
$$y_1 := g_s.$$

Proof. In fact, applying this method to the Dahlquist test equation, we obtain

$$g_0 := y_0, \quad g_1 := y_0 + zg_0,$$
$$g_i := (2/s^2)zg_{i-1} + 2g_{i-1} - g_{i-2} = 2\left(1 + z/s^2\right)g_{i-1} - g_{i-2}, \quad (4.98)$$

which is exactly the same recursion formula for the Chebyshev polynomials in the point $x = 1 + z/s^2$. □

Actually in the points $T_s(1 + z/s^2) = \pm 1$ there is no damping at all of the higher frequencies and the stability domain has zero width. We therefore choose a small $\epsilon > 0$, for example $\epsilon = 0.05$ and put

$$R_s(z) = \frac{1}{T_s(w_0)}T_s(w_0 + w_1 z), \quad w_0 = 1 + \frac{\epsilon}{s^2}, \quad w_1 = \frac{T_s(w_0)}{T'_s(w_0)} \quad (4.99)$$

These polynomials oscillate between $1 - \epsilon$ and $-1 + \epsilon$. The stability domains become a bit shorter (by $4\epsilon s^2/3$), but the boundary is in a safe distance form the real axis [24]. Let's calculate the Runge-Kutta method we need in this case.

$$T_s(w_0 + W_1) = 2w_0 T_{s-1}(w_0 + w_1 z) + 2w_1 z T_{s-1}(w_0 + w_1 z) - T_{s-2}(w_0 + w_1 z) \quad (4.100)$$

For the definition of R_s

$$R_s(z) = \frac{1}{T_s(w_0)}\left(2w_0 R_{s-1}(z)T_{s-1}(w_0) + 2w_1 z R_{s-1}(z)T_{s-1}(w_0) - T_{s-2}(w_0)\right), \quad (4.101)$$

which leads to the method

$$g_0 := y_0, \quad g_1 := y_0 + \Delta t(w_1/w_0)f(g_0),$$
$$g_i := (1/T_i(w_0))\left(2w_1\Delta t T_{i-1}(w_0)f(g_{i-1}) + 2w_0 T_{i-1}(w_0)g_{i-1} - T_{i-2}(w_0)g_{i-2}\right). \quad (4.102)$$

Here we presented only the first order RK-Chebyshev methods because higher order methods are much more complicated to implement. Anyway we are not aiming for very high accuracy because we are mainly interested in the final state and not in the transients and we are only comparing things statistically; so high accuracy is not really needed here.

4.3.3. Numerical simulations

In chapter 3 we reported a set of standard parameters which fitted with our experiments. We recall them here:

1	2	1	2	2		K	L_x [nm]	L_y [nm]
0.025	0.025	0.0005	0.0005	20^0	0.01	1.2	40000	3.75	2.75

Table 4.1.: Summary of the parameters used in the simulations.

At confluence we did not observe any aggregations, but a pattern formation in angle, as can be seen in Fig. 4.7. In this section we would like to analyze the eect of the dierent parameters on the pattern formation systematically, varying one parameter at a time. We perform the same simulations as before, starting with random initial conditions and show the results at the last day of culture. Since the colorbar always remains the same, we omit it for ease of readibility in subsequent pictures.

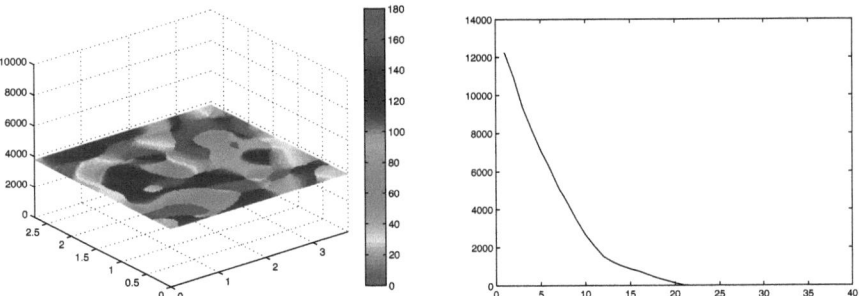

Figure 4.7.: *Simulation at day 9 with the standard parameters reported in Table 4.1.*

Diffusion coefficients

Both diffusion coefficients are responsible for the formation of angular patterns at confluence. An increase in the angular coefficient leads to alwasy narrower patterns until a uniform distribution in angle occurs. If the spatial coefficient increases, the number of winning directions decreases until only one direction wins, that is all the cells are oriented in the same direction (Fig. 4.8, 4.9, 4.10).

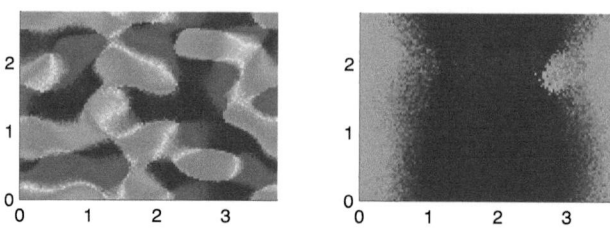

Figure 4.8.: *On the left we doubled the angular coefficient ($\sigma_1 = 0.05$) and on the right the spatial coefficient ($\sigma_2 = 0.05$).*

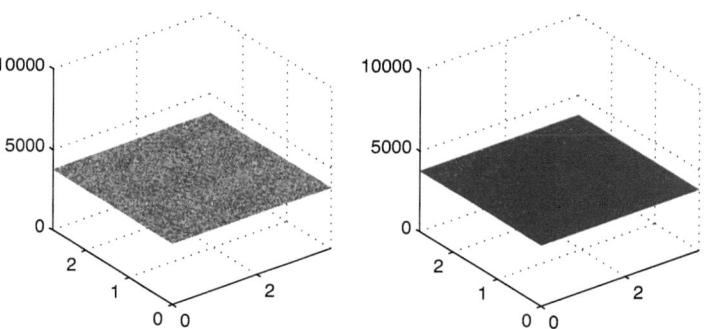

Figure 4.9.: *In the left frame the angular coefficient is ten times larger than the standard one ($\sigma_1 = 0.25$), whereas in the right one is the spatial coefficient is ten times larger ($\sigma_2 = 0.25$).*

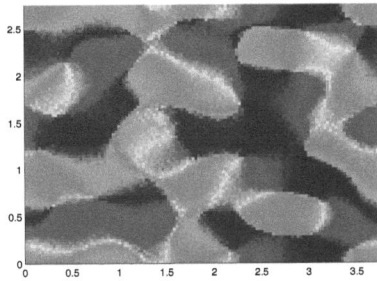 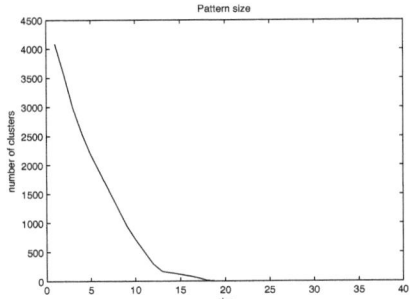

Figure 4.10.: *Doubling the angular coefficient ($\beta_1 = 0.05$) we observe a slight decrease of the pattern size.*

Drift parameters

The drift parameters have an influence on the size of the patterns as well as on the spatial stability of the model. A larger angular drift results in larger patterns (Fig. 4.11), whereas a larger spatial coefficient leads to narrower pattern and aggregation (Fig. 4.13).

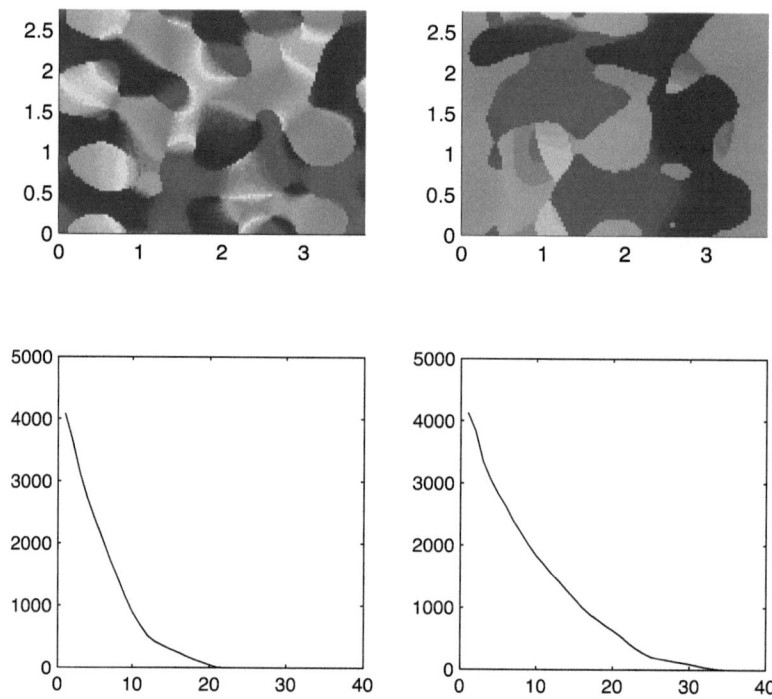

Figure 4.11.: *Doubling the angular drift coefficient from $\epsilon_1 = 0.0005$ to $\epsilon_1 = 0.001$ we observe an enlargement of the pattern size.*

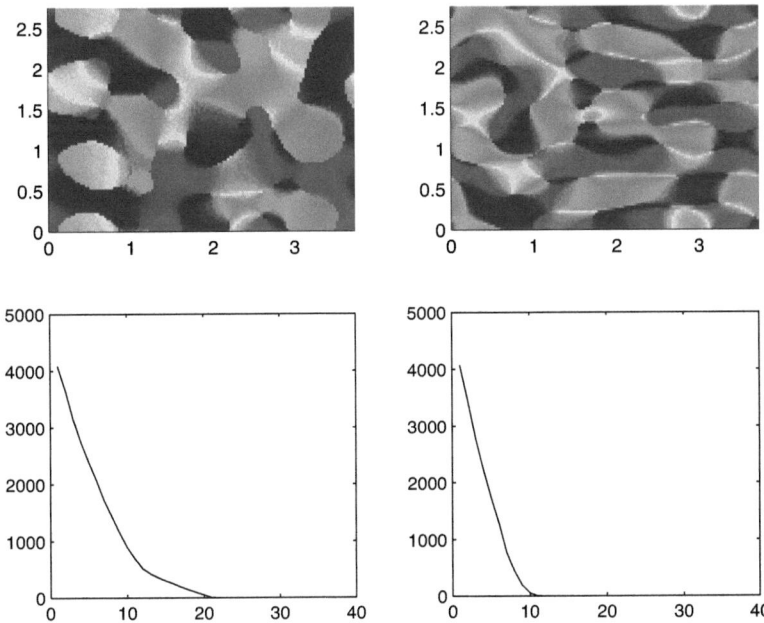

Figure 4.12.: *Comparing $\epsilon_2 = 0.0005$ with $\epsilon_2 = 0.0006$ we can observe the difference in pattern size.*

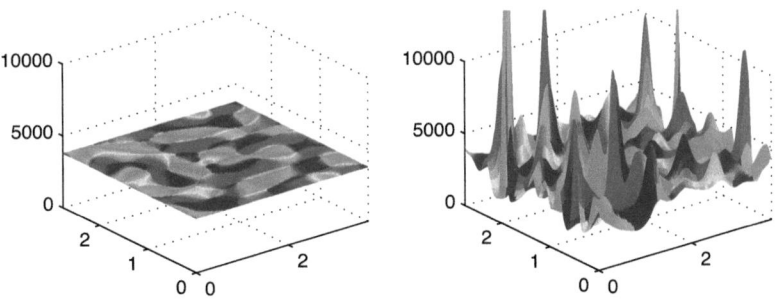

Figure 4.13.: *Further increase in $\epsilon_2 = 0.0006$ to $\epsilon_2 = 0.0007$ leads to instabilities in space.*

5. Conclusions

We started from a relativly simple model consisting of an ordinary dierential equation to simulate the number of cells in time. This model was then extended to capture the non-instantaneous growth of chondrocytes and the tendency of the growth to slow down with time. These adaptations gave us interesting results and gave also explanations on the behavior of the cells in two dierent conditions.
In particular, we were interested on the two-dimensional growth of cells in presence of growth factors, condition in which the chondrocytes are very similar to fibroblasts. This allowed us to start with an already developed model.

We extended the nonlinear spatio-angular model, by including cell duplication, and performed simulations in three dimensions (two for the space and one for the angle). However, we needed a comparison between the numerical results and the experimental data. For this reason, we performed specific experiments to get some of the parameters and fitted others with the model. An important tool were Gabor filters, through which we tranformed the cell pictures in colored pictures of the same type as that obtained with the simulations. The estimated parameters may be used to measure behaviors of cells from dierent donors during ex vivo culture to optimize the expansion conditions for clinical application.

With this study we showed that there is a potentiality to use mathematics in the life science, in particular it is possible to apply mathematics to biology. There are still a lot of open problems, but the science is never a process with an end.

APPENDIX

A. Logistic equation

A.1. Classical logistic equation

Verhulst (1838) proposed the inclusion of a self-limiting process to adjust a simple exponential model $dy/dt = \rho y(t)$, suggesting the equation

$$\dot{y}(t) = \rho y(t) \left(1 - \frac{y(t)}{K}\right), \tag{A.1}$$

where K is the carrying capacity of the environment and ρ the growth rate. This equation has two steady states, $y = 0$ and $y = K$. For $\rho > 0$ the first one is unstable whereas the second one is stable. We can calculate the exact solution of this equation by separation of variables:

$$-\ln\left(\frac{1 - \frac{y}{K}}{y}\right) = \rho t + C_1, \tag{A.2}$$

for some constant C_1 determined by the initial conditions. This leads to the solution

$$\overline{y}(t) = \frac{K}{1 + KC_1 e^{-t}}, \tag{A.3}$$

where $C_1 = \dfrac{1}{y_0} - \dfrac{1}{K}$ for $y(0) = y_0$ [37]. If $y_0 < K$, $y(t)$ simply increases monotonically to K, while if $y_0 > K$ it decreases monotonically to K. In the former case there is a qualitative di erence dependi ng on whether $y_0 > K/2$ or $y_0 < K/2$ (see Fig. A.1).

A.2. Delay logistic equation

One of the deficiencies of single population models is that the birth rate is considered to act instantaneously whereas there may be a time delay to take account of the time to reach maturity, the finite gestation period and so on. When we add such a discrete delay T in the logistic equation (A.1), it is necessary to define a fictitious function $\psi(t)$ that determines the rate at which new cells appear over $[0, T)$. However, one should not interpret $\psi(t)$ as the number of cells $y(t)$ for negative t in $[-T, 0)$, but rather $y(t - T)$ as the rate of the cell growth for t in $[0, T)$. If the growth is synchronous and the cells divide around some specific time, $\psi(t)$ could be a Gaussian centered about that time, but if the growth is asynchronous, $\psi(t)$ should be a constant. In either case, the function is normalized by assuming that the number of cells duplicates over the first interval $[0, T)$. In an asynchronous growth, this condition lead to the choice of

$$\psi(t) = \frac{y_0}{\rho T}, \quad t \in [-T, 0). \tag{A.4}$$

We can incorporate the delay in two ways. We will analyze both in details, but it is the first one that fitted our experiments. In both cases we have the same steady states as in the former equation (A.1). We will in particular analyze the stability of the state $y(t) = K$, linearizing around it. We consider a perturbation $z(t)$ of the steady state

$$y(t) = K + \epsilon z(t). \tag{A.5}$$

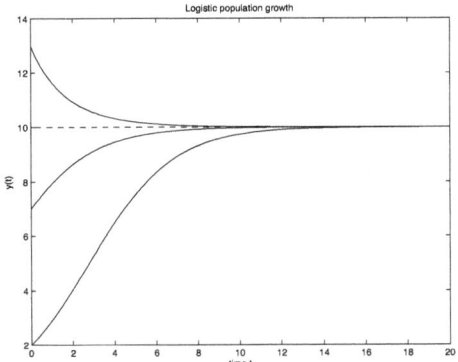

Figure A.1.: *Logistic equation with = 0.5, K = 10. For different initial values we have a different behavior of the solution which tends to the stable steady state y(t) = K.*

A.2.1. Delay in the linear term

$$\dot{y}(t) = \rho y(t-T)\left(1 - \frac{y(t)}{K}\right). \quad (A.6)$$

Substituting (A.5) in (A.6) we get

$$\epsilon \dot{z}(t) = \rho(\epsilon z(t-T) + K)\left(-\epsilon \frac{z(t)}{K}\right). \quad (A.7)$$

If we neglect the second order terms in ϵ, we have the equation for the perturbation $z(t)$

$$\dot{z}(t) = -\rho z(t). \quad (A.8)$$

Such perturbation always decreases for positive ρ. The steady state is then stable independently from T.

A.2.2. Delay in the quadratic term

$$\dot{y}(t) = \rho y(t)\left(1 - \frac{y(t-T)}{K}\right). \quad (A.9)$$

With the same method as before we obtain for the perturbation

$$\dot{z}(t) = -\rho z(t-T), \quad (A.10)$$

which needs a more careful analysis. Assuming that the solutions are exponential, e^t, we obtain the characteristic equation

$$\lambda + \rho e^{-T} = 0. \quad (A.11)$$

We wish to determine the stability boundaries, namely those values of the parameter T, for which the real part of a root of this equation is zero. Let us then consider the pure imaginary $\lambda = if$. Changes in

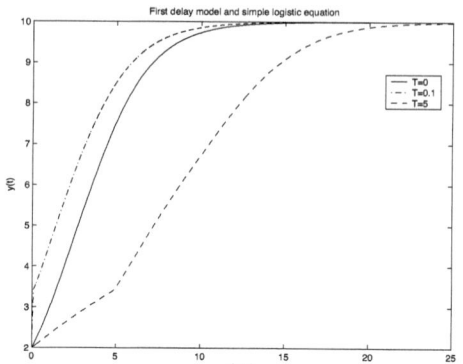

Figure A.2.: *Logistic delay equation with $\rho = 0.5$, $K = 10$ and different delays. The steady state $y(t) = K$ is stable for any delays, but the qualitative behavior is different from the simple logistic equation.*

stability occur at the intersections between the unit circle e^{-ifT} and the delay curve $-if/\rho$. We have exactly two intersections, which can be calculated (see also Fig. (A.3)).

$$if = -\rho e^{-ifT_1} = -\rho[\cos(fT_1) - i\sin(fT_1)] \qquad (A.12)$$

If we set equal the real and imaginary parts, we obtain the equations

$$\cos(fT_1) = 0 \qquad (A.13)$$
$$f = \rho \sin(fT_1), \qquad (A.14)$$
$$(A.15)$$

so that

$$fT_1 = k\frac{\pi}{2}, \quad k \in \mathbb{Z} \qquad (A.16)$$

As we assume the delay to be positive, the critical delay is given by

$$T_1 = \frac{\pi}{2\rho}. \qquad (A.17)$$

For any $T > \pi/(2\rho)$ the steady state will not be stable anymore. In the next pictures we observe that for small delays though oscillations occur we finally reach the steady state. On the contrary, if the delay is large enough we have periodic oscillations.

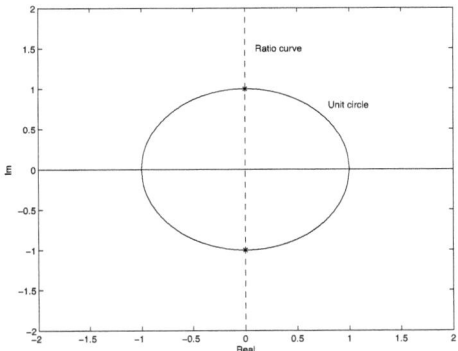

Figure A.3.: *Ratio curve and unit circle.*

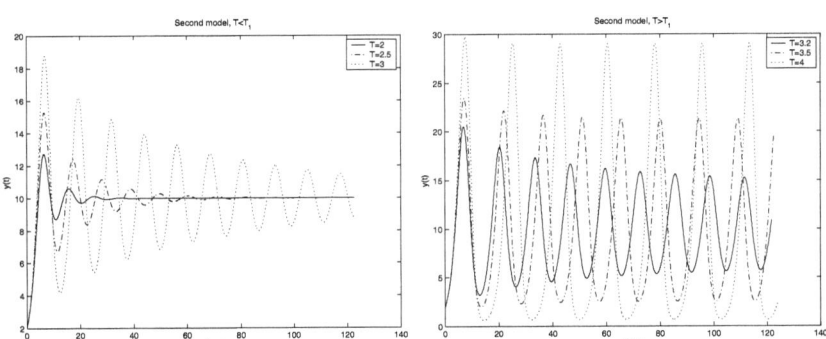

Figure A.4.: *Logistic delay equation, = 0.5. The steady state $y(t) = K$ is stable for small value of T. As the delay increases, reaching the critical delay $T_1 =$, the steady state starts to destabilize. The steady state is unstable for large value of T. The more the delay increases, the more the model is.*

B. Spatio-angular model, calculations and inequalities

B.1. Governing Equation

Let start with the problem in space. Because we study a well with several cells growing, the cell population can be considered as a continuum substance, with a mass density $C(\vec{x}, t)$, immersed in a fictitious fluid which fills the region D. Let $\vec{x}$ be a point in D and $\vec{u}(\vec{x}, t)$ the velocity of the particle of this substance that is moving through $\vec{x}$ at time t. Then, the mass of substance in each subregion $U \subset D$ at time t is given by

$$m(U, t) = \int_U C(\vec{x}, t) dx. \tag{B.1}$$

If mass is neither created nor destroyed the rate of change of mass in U is

$$\frac{d}{dt} m(U, t) = \frac{d}{dt} \int_U C(\vec{x}, t) dx \tag{B.2}$$

Now, if ∂U is the boundary of U, $\vec{n}$ the unit outward normal defined at points of ∂U, the mass flow rate per unit area is $\vec{F}(C) \cdot \vec{n}$, where $\vec{F}$ is the flux. Then

$$\frac{d}{dt} \int_U C(\vec{x}, t) dV = - \int_{\partial U} \vec{F}(C) \cdot \vec{n} \, dA. \tag{B.3}$$

By the divergence theorem and considering that U does not change in time we obtain the integral form of the law of conservation of mass

$$\int_U \left[\frac{\partial C}{\partial t} + \operatorname{div}(\vec{F}(C)) \right] dV = 0. \tag{B.4}$$

Because this is to hold for all U, it is equivalent to the continuity equation

$$\frac{\partial C}{\partial t} + \operatorname{div}(\vec{F}(C)) = 0. \tag{B.5}$$

Without considering the diusion, the mass flow rate across ∂U per unit area would be $C\vec{u} \cdot \vec{n}$. If we add the diusion term we obtain $\vec{F}(C) = C\vec{u} - \epsilon \, C$ where ϵ is the diusivity [2]. Now, we substitute the form of $\vec{F}(C)$ in (B.5) to have

$$\frac{\partial C}{\partial t} + \operatorname{div}(C\vec{u}) = \epsilon \Delta C. \tag{B.6}$$

What remains to be defined is the particular form that $\vec{u}$ assumes in our case. Very small cells here move in highly viscous media, so that we can assume that the velocity is proportional to a corresponding conservative force. Moreover, the force is the derivative of some potential function P: a given cell is subject to the eect of a potential that represents its cumul ative interaction with the other cells. Then it can be written in the form $P = W * C$, where

$$(W * C)(\vec{x}, \theta, t) = \int_- \int_D W(\vec{x} - \vec{x'}, \theta, \theta') \cdot C(\vec{x'}, \theta', t) dx' d\theta' \tag{B.7}$$

represents the influence of the distribution of cell $C(\vec{x}', \theta', t)$ on the angle θ at position $\vec{x}$. Thus, a term of the form $W * C$ represents the influence of the cell distribution at angle θ and position $\vec{x}$; the term $C(\vec{x}, \theta)(W * C)$ is the rate at which cells contact, align and aggregate to other cells at angle θ and position $\vec{x}$. If we call γ the constant of proportionality, we obtain then the equation

$$\frac{\partial C}{\partial t} + \text{div}(C\gamma \ (W * C)) = \epsilon \Delta C. \tag{B.8}$$

In particular, considering that C depends also on the angle of orientation θ, our model can be written as

$$\frac{\partial C}{\partial t} = \epsilon_1 \frac{\partial^2 C}{\partial \theta^2} + \epsilon_2 \Delta_{xy} C - \gamma_1 \frac{\partial}{\partial \theta} \left(C \frac{\partial}{\partial \theta} [W * C] \right) - \gamma_2 \ _{xy} (C \ _{xy} [W * C]), \tag{B.9}$$

where ϵ_1, ϵ_2 are respectively the angular and spatial diusion coecients and γ_1, γ_2 the angular and spatial drift coecients. This equation describes the convectional drift of the cells in physical and angular space towards the points of highest concentration, causing alignment and aggregation.

B.2. Normalization of the kernels

We want to normalize the function

$$W(x) = e^{-\frac{x^2}{2\sigma^2}} \quad x \in [0, L]. \tag{B.10}$$

We can calculate the integral of $W(x)$ on $[0, L]$ considering that

$$\left(\int_0^\infty e^{-x^2} dx \right)^2 = \int_0^\infty \int_{-\infty}^\infty e^{-\frac{(x^2+y^2)}{2}} dx\, dy. \tag{B.11}$$

If we apply a transformation in polar coordinates, we find:

$$\int_{-\infty}^0 e^{-x^2} dx = \frac{\sqrt{\pi}}{2}, \quad \int_0^\infty e^{-\frac{x^2}{2\sigma^2}} dx = \sigma \sqrt{\frac{\pi}{2}}. \tag{B.12}$$

That means, to have a normalized and periodic kernel, we can choose

$$W(x) = \frac{1}{\sigma\sqrt{2\pi}} e^{-\frac{x^2}{2\sigma^2}}. \tag{B.13}$$

B.3. Inequalities

To obtain our estimates we need the following inequalities that we present for the spaces which we use.

Inequality B.3.1 (Cauchy). *Let $a, b > 0$, then*

$$ab \leq \frac{1}{2}(a^2 + b^2) \tag{B.14}$$

Proof.

$$0 \leq (a-b)^2 = a^2 - 2ab + b^2.$$

□

Inequality B.3.2 (Hölder). *Assume $u \in L^p(U)$ and $v \in L^q(U)$, with $1 \leq p, q \leq \infty$, $\frac{1}{p} + \frac{1}{q} = 1$ then*

$$\int_U |uv|\, dx \leq \|u\|_{L^p(U)} \|v\|_{L^q(U)}. \tag{B.15}$$

Proof. By homogeneity, we may assume $\|u\|_{L^p(U)} = \|v\|_{L^q(U)} = 1$. Then Cauchy's inequality implies that

$$\int_U |uv|\, dx \leq \frac{1}{p}\int_U |u|^p dx + \frac{1}{q}\int_U |v|^q dx = 1 = \|u\|_{L^p(U)} \|v\|_{L^q(U)}.$$

□

Inequality B.3.3 (Cauchy, with δ).

$$ab \leq \delta a^2 + \frac{b^2}{4\delta} \quad (a, b > 0, \delta > 0). \tag{B.16}$$

Proof. Apply Cauchy's inequality to

$$ab = ((2\delta)^{\frac{1}{2}} a)(b(2\delta)^{-\frac{1}{2}}).$$

□

Inequality B.3.4 (Gronwall). *Let $\eta(.)$ be a nonnegative, absolutely continuous function on $[0, T]$, which satisfies for a.e. t the differential inequality*

$$\eta'(t) \leq \phi(t)\eta(t) + \psi(t), \tag{B.17}$$

where $\phi(t)$ and $\psi(t)$ are nonnegative, summable functions on $[0, T]$. Then

$$\eta(t) \leq e^{\int_0^t \phi(s)ds}\left[\eta(0) + \int_0^t \psi(s)ds\right], \tag{B.18}$$

for all $0 \leq t \leq T$.

Proof. From (B.17) we see

$$\frac{d}{ds}\left(\eta(s)e^{-\int_0^s \phi(r)dr}\right) = e^{-\int_0^s \phi(r)dr}(\eta'(s) - \phi(s)\eta(s)) \leq e^{-\int_0^s \phi(r)dr}\psi(s)$$

for a.e. $0 \leq s \leq T$. Consequently for each $0 \leq t \leq T$, we have

$$\eta(t)e^{-\int_0^t \phi(r)dr} \leq \eta(0) + \int_0^t e^{-\int_0^s \phi(r)dr}\psi(s)ds \leq \eta(0) + \int_0^t \psi(s)ds.$$

□

C. Spatial diffusion

C.1. Random walks

If we study the movement of the cells in space we can consider each cell as a "walker" which perform a random walk. Random walks have been widely studied and in the next section we would like to summarize this theory before using it.

A random walk considers a "walker" which starts somewhere, and takes steps in a random direction. In some cases the step can be of random length as well. In the limit as the step length and the time between steps go to zero, the random walker typically exhibits a form of Brownian motion. Let's first model Brownian motion as a sum of independent random displacements in one direction. Imagine the Brownian particle starts at the origin $x = 0$ and is free to move in either direction along the x-axis. The net effect of many individual molecular impacts is to displace the particle a random amount X_i in each interval of duration Δt. Assume each displacement X_i realizes one of two possibilities, $X_i = x_1 = +\Delta x$ or $X_i = x_2 = -\Delta x$, with equal probabilities $P(x_j) = 1/2, j = 1, 2$ and that the variables X_i are statistically independent (the outcome of each X_i does not depend on the outcome of the other X_i). After n such intervals the net displacement X is $X = X_1 + X_2 + ... + X_n$. This is the random step or random walk model of Brownian motion. The mean of a random variable X is defined as $<X> = \sum_j x_j P(x_j)$ where the sum is over all possible realizations x_j of X. Because the X_i are statistically independent through the variance sum theorem (the variance of the sum is equal to the sum of the variances), since the total duration of the walk is $t = n\Delta t$ we get

$$<X^2> = \left(\frac{\Delta x^2}{\Delta t}\right) t. \tag{C.1}$$

This equation expresses the signature property of Brownian motion: the variance $<X^2>$ of the net displacement X is proportional to the time t during which that displacement is made [32]. Actually the ratio $\Delta x^2/\Delta t$ is a physically meaningful constant, equal twice the diffusion constant D. To see this, let's consider the diffusion equation

$$\frac{\partial C(x,t)}{\partial t} = D \frac{\partial^2 C(x,t)}{\partial x^2}, \tag{C.2}$$

where $C(x, t)$ is the particle number density (particle number per unit area) at position x and time t. Integrating, we find the solution

$$C(x,t) = \sqrt{\frac{1}{4\pi D t}} e^{-\frac{x^2}{4Dt}}. \tag{C.3}$$

If we want to calculate again the $<X^2>$, given in this case by

$$<X^2> = \frac{\int_{-\infty}^{+\infty} x^2 C(x,t) dx}{\int_{-\infty}^{+\infty} C(x,t) dx}, \tag{C.4}$$

using the following two expressions

$$\int_0^\infty x^2 e^{-ax^2} dx = \frac{1}{4}\sqrt{\frac{\pi}{a^3}} \quad \int_0^\infty e^{-ax^2} dx = \frac{1}{2}\sqrt{\frac{\pi}{a}}, \tag{C.5}$$

we obtain
$$<X^2> = 2Dt. \qquad (C.6)$$
It is easy to generalize it to the two dimensional case, substituting the displacement Δx with $\sqrt{\Delta x^2 + \Delta y^2}$.

C.2. Experiments

We perform experiments to approximate the spatial diffusion constant. With a time-laps microscope (OLYMPUS IX81) we obtain a sequence of pictures in time which we analyze with a manual tracking (Software:analySIS^D). The cells are cultured in presence of growth factors in the same way as described in [6]. In some previous experiments (Fig. C.1) we observed that a cell moves with a mean velocity of

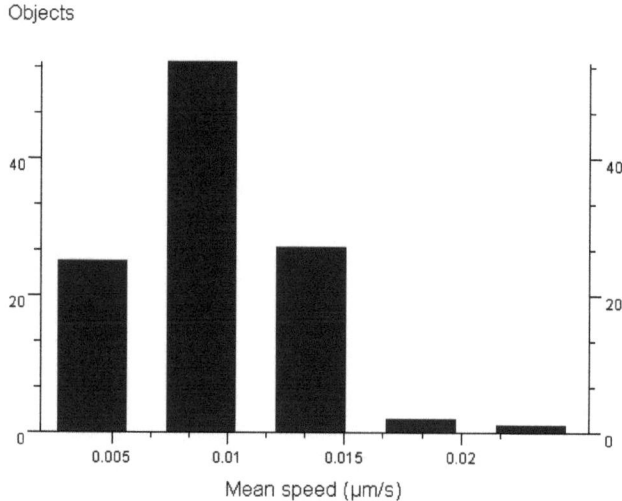

Figure C.1.: *Mean speed of the cells.*

$0.6\mu m$ pro minute, making then $9\mu m$ in 15 minutes. As the cell has a length of about $50\mu m$ with a sequence of frames taken every 15 minutes we can capture the meaningful movement of a cell. With a manual tracking (Fig. 3.2, section 3.2.1) we can follow each cell for 12 hours, a duration that allowed us to neglect the doubling of the cells. At each time t we calculate the mean of the square of the displacement and fit the resulting points with a straight line ($<X^2> = 2Dt$, Fig. C.2). The slope of this line divided by 2 give us exactly the diffusion coefficient. We make experiments at different densities (200, 3000, 10000, 15000, 20000 cells/scm) and for each density one hundred cells in two or three zones of the well are followed. In table (C.1) the diffusion coefficients are reported. It seems that the coefficient initially tends to increase with the density to decrease again for higher densities. Actually, for densities higher

cells/scm	200	3000	10000	15000	20000
coe D1	0.31	0.30	0.37	0.31	0.31
coe D2	0.18	0.23	0.32	0.30	0.26
coe D3	0.23		0.40	0.34	0.30
mean ± SD	0.24 ± 0.07	0.26 ± 0.04	0.36 ± 0.04	0.31 ± 0.02	0.30 ± 0.03

Table C.1.: Summary of the results of experiments at dierent densities and mean values ± SD.

then 200 we can think to fit the variance with αt. Beta equal one is normal diusion; beta larger than one is called super-diusion. This is the case where all random walkers are moving away from each other at constant time. Beta equal two corresponds to the ballistic motion, such as the particles of a bomb which explodes. Again for the density 10000 this exponent is the highest. It could be considered the

cells/scm	3000	10000	15000	20000
1	1.11	1.44	1.18	1.18
2	1.09	1.24	1.19	1.07
3		1.26	1.15	1.00
mean	1.1	1.31	1.17	1.08

Table C.2.: Summary of the results of experiments at dierent densities and mean values.

most suitable condition for the cells to move: they repel themselves and still have enough space to move. Indeed in our experiments the mean exponent is 1.16, so that we can maybe conclude that the repulsion is negligible for these densities and consider $\beta = 1$. Then we find that the spatial coecient can be approximated by the value $D = 0.29\,\mu m^2/s$. As we perform all the experiments in days and mm^2, we need to find how large is the coecient in the new unit of measure. Let's consider the heat equation

$$u_t = D u_{xx}, \qquad (C.7)$$

where u is the cell density, u_t the first derivative in time and u_{xx} the second derivative in space. Let $u^*(x^*, t^*) = u(x, t)$, with the non dimensional variables $x^* = \frac{x}{L}$ and $t^* = \frac{t}{T}$, where $L = 1000\,\mu m^2$ and $T = 1\,day = 86400s$. Calculating the derivatives we obtain

$$u_t = u_{t^*}^* \frac{dt^*}{dt} = u_{t^*}^* \frac{1}{T} \qquad (C.8)$$

$$u_{xx} = u_{x^*x^*}^* \left(\frac{dx^*}{dx}\right)^2 = u_{x^*x^*}^* \frac{1}{L^2}. \qquad (C.9)$$

Substituting it in the heat equation (C.7) we obtain the dimensionless equation

$$u^*_{t^*} = D^* u^*_{x^*x^*}, \qquad (C.10)$$

where

$$D^* = \frac{D^* T}{L^2} = 0.025. \qquad (C.11)$$

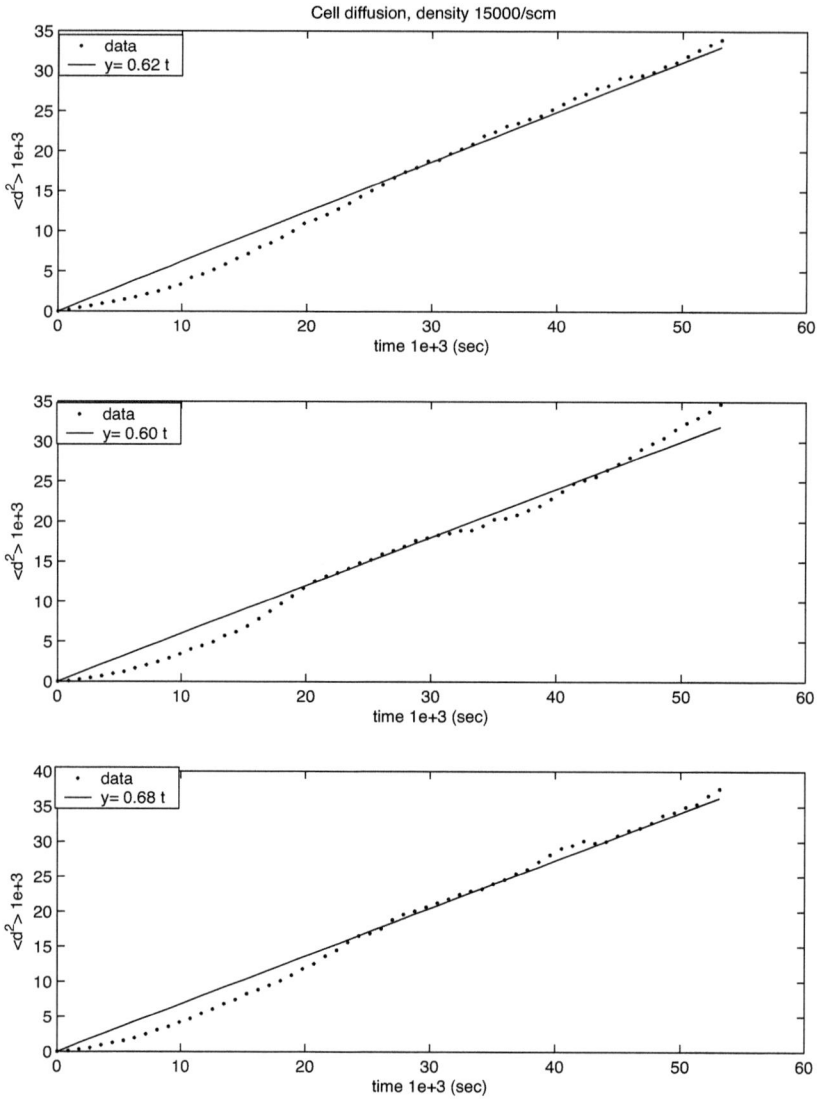

Figure C.2.: *Fitting of the diffusion constant with a straight line for a density of 15000 cells/scm.*

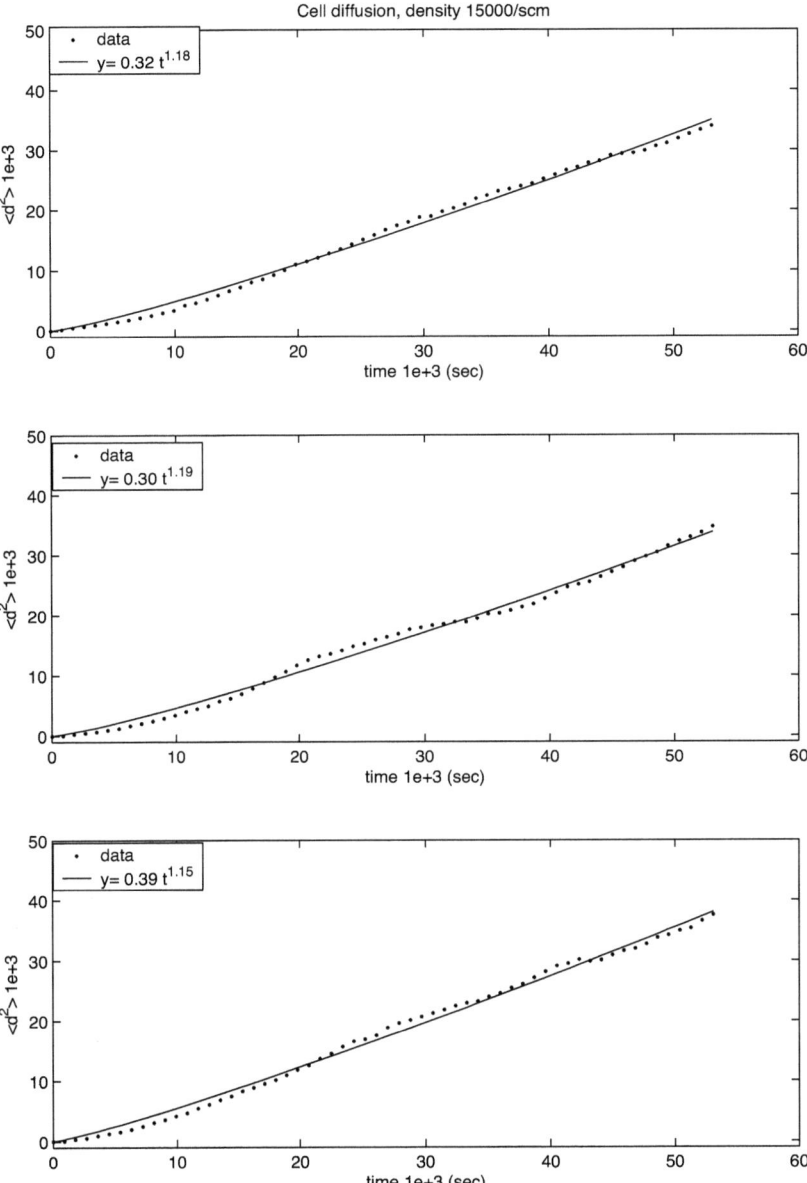

Figure C.3.: *Fitting of the diffusion constant with t for a density of 15000 cells/scm.*

Bibliography

[1] Abramowitz M., S.I.A.: Handbook of Mathematical Functions with Formulas, Graphs, and Mathematical Tables. Dover Publications (1964)

[2] A.Chorin, Marsden, J.: A Mathematical Introduction to Fluid Mechanics. Springer-Verlag (2000)

[3] A.Gouillou, Lago, B.: Domaine de stabilité associé aux formules d'intégration numérique d'équations diérentielles, a pas séparés et a pas liés. recherche d e formules a grand rayon de stabilité. Ier Congr. Ass. Fran. Calcul., AFCAL pp. 43–56 (1960)

[4] Baker, C., Bocharov, G., Paul, C., Rihan, F.: Modelling and analysis of time-lags in some basic patterns of cell proliferation. J Math Biol **37**, 341–371 (1998)

[5] Barbero, A., Grogan, S., Schafer, D., Heberer, M., Mainil-Varlet, P., Martin, I.: Age related changes in human articular chondrocyte yield, proliferation and post-expansion chondrogenic capacity. Ostearthritis Cartilage **12**, 476–484 (2004)

[6] Barbero, A., Palumberi, V., Wagner, B., Sader, R., Grote, M.J., Martin, I.: Experimental and mathematical study of the influence of growth factors on the growth kinetics of adult human articular chondrocytes. Journal of Cellular Physiology **204**, 830–838 (2005)

[7] Barbero, A., Ploegert, S., Heberer, M., Martin, I.: Plasticity of clonal populations of dedierentiated adult human articular chondrocytes. Arthritis Rheum **48**, 1315–1325 (2003)

[8] Beattie, G., Cirulli, V., Lopez, A., Hayek, A.: Ex vivo expansion of human pancreatic endocrine cells. J Clin Endocrinol Metab **82**, 1852–1856 (1997)

[9] Carpenter, M., Cui, X., Hu, Z., Jackson, J., Sherman, S., Seiger, A., Wahlberg, L.: In vitro expansion of a multipotent population of human neural progenitor cells. Exp Neurol **158**, 265–278 (1999)

[10] Chipot, M., Edelstein-Keshet., L.: A mathematical theory of size distribution in tissue culture. J. Math. Biol. **16**, 115–130 (1983)

[11] Civelecoglu, G., Edelstein-Keshet, L.: Modelling the dynamics of f-actin in the cell. Bull. Math. Biol. **56**, 587–616 (1994)

[12] Comincioli, V.: Analisi numerica. McGraw-Hill (1995)

[13] Daugman, J.: Uncertainty relations for resolution in space, spatial frequency, and orientation optimized by two-dimensional visual cortical filters. Journal of the Optical Society of America A **2**, 1160–1169 (1985)

[14] Deasy, B., Qu-Peterson, Z., Greenberger, J., Huard, J.: Mechanisms of muscle stem cell expansion with cytokines. Stem Cells **20**, 50–60 (2002)

[15] Deenick, E., Gett, A., Hodgkin, P.: Stochastic model of T dell proliferation: A calculus revealing il-2 regulation of precursor frequencies, cell cycle time, and survival. J. Immunol. **170**, 4963–4972 (2003)

[16] Edelstein-Keshet, L., Ermentrout, G.: Models for contact-mediated pattern formation: cells that form parallel arrays. J. Math. Biol. **29**, 3–58 (1990)

[17] Elsdale, T.: The generation and maintenance of parallel arrays in cultures of diploid fibroblasts. E. Kulonen and J. Pikkarainen (eds), Biology of Fibroblast pp. 41–58 (1973)

[18] Elsdale, T., Bard, J.: Collagen substrata for studies on cell behaviours. Journal of Cell Biology **54**, 626–637 (1972)

[19] Erickson, C.: Analysis of the formation of parallel arrays in bhk cells in vitro. Exp. Cell Res. **115**, 303–315 (1978)

[20] Evans, L.C.: Partial Dierential Equations. American Mathematical Society (1998)

[21] Forsyth, D.A., Ponce, J.: Computer Vision, A Modern Approach. Prentice Hall (2003)

[22] Franklin, J.: Numerical stability in digital and anlogue computation for diusion problems. J.Math.Phys. **37**, 305–315 (1959)

[23] Gabor, D.: Theory of comminication. J. IEE **93**(26), 429–457 (1946)

[24] Hairer E., W.G.: Solving Ordinary Dierential Equatio n II. Springer-Verlag (2002)

[25] Hairer E. Noersett S.P., W.G.: Solving Ordinary Diere ntial Equations. I: Nonsti Problems. Springer-Verlag (1987)

[26] Harley, C., Futcher, A., Greider, C.: Telomeres shorten during ageing of human fibroblasts. Nature **345**, 458–460 (1990)

[27] Israel, G., Gasca, A.M.: The Biology of Numbers. Birkhäuser (2002)

[28] Jakob, M., Demarteau, O., Schafer, D., Hinterman, B., Dick, W., Heberer, M., Martin, I.: Specific growth factors during the expansion and redierentiation o f adult human articular chondrocytes enhance chondrogenesis and cartilaginous tissue formation in vitro. J Cell Biochem **81**, 368–377 (2001)

[29] Kelley, C.: Solving Nonlinear Equations with Newton's Method. Siam (2003)

[30] Kress, R.: Linear Integral Equations. Springer-Verlag (1999)

[31] Langer, R., Vacanti, J.: Tissue engineering. Science **260**, 920–926 (1993)

[32] Lemons, D.S.: An Introduction to Stochastic Processes in Physics. Johns Hopkins Paperback (2002)

[33] Marko, C.L.: Ueber polynome, die in einem gegebenen in tervall möglichst wenig von null abweichen. Math. Ann. **77**, 213–258 (1916)

[34] Mogilner, A., Edelstein-Keshet, L.: Selecting a common direction, how orientational order can arise from simple contact responses between interacting cells. J. Math. Biol. **33**, 619–660 (1995)

[35] Mogilner, A., Edelstein-Keshet, L.: Spatio-angular order in populations of self-aligning objects: formation of oriented patches. Physica D **89**, 346–367 (1996)

[36] Mogilner, A., Edelstein-Keshet, L., Ermentrout., G.B.: Selecting a common direction. ii. peak-like solutions representing total alignment of cell clusters. J. Math. Biol. **34**, 811–842 (1996)

[37] Murray, J.: Mathematical Biology. Springer-Verlag (2001)

[38] Oster, G.F., Murray, J.D., Harris, A.K.: Mechanical aspects of mesenchymal morphogenesis. J. Embryol. exp. Morph. **78**, 83–125 (1983)

[39] Pittelkow, M., Cook, P., Shipley, G., Derynck, R., Coe y, R.: Autonomous growth of human keratinocytes requires epidermal growth factor receptor occupancy. Cell Growth Dier **4**, 513–521 (1993)

[40] Sharma, B., Elissee, J.H.: Engineering structurally organized cartilage and bone tissues. Annals of biomedical enginnering **32**, 148–159 (2004)

[41] Sherley, J., Stadler, P., Johnson, D.: Expression of the wild-type p53 antioncogene induces guanine nucleotide-dependent stem cell division kinetics. Proc Natl Acad Sci **92**, 136–140 (1995)

[42] Sherley, J., Stadler, P., Stadler, J.: A quantitative method for the analysis of mammalian cell proliferation in culture in terms of dividing and non-dividing cells. Cell Prolif **28**, 137–144 (1995)

[43] Simmons, P., Haylock, D.: Use of hematopoietic growth factors for in vitro expansion of precursor cell populations. Curr Opin Hematol **2**, 189–195 (1995)

[44] Smith, G.: Numerical Solution of Partial Dierential E quations: Finite Dierence Methods. Oxford University Press (1986)

[45] Stewart, J., Masi, T., Cumming, A., Molnar, G., Wentworth, B., Sampath, K., McPherson, J., Yaeger, P.: Characterization of proliferating human skeletal muscle-derived cells in vitro: dierential modulation of myoblast markers by tgf-beta2. J Cell Physiol **196**, 70–78 (2003)

[46] Thomas, J.W.: Numerical Partial Dierential Equation s. Springer-Verlag (1998)

[47] Trefethen, L., Trefethen, A., Reddy, S., Driscoll, D.: Hydrodynamics without eigenvalues. Science **261**, 578 (1993)

[48] Trinkaus, J.P.: Further thoughts on directional cell-movement during morphogenesis. Journal of Neuroscience research **13**, 1–19 (1985)

[49] Volterra, V.: Variazioni e fluttuazioni del numero d'individui in specie animali conviventi. Memorie della R. Accademia dei Lincei **2**, 31–113 (1926)

[50] Werner, D.: Funktionalanalysis. Springer-Verlag (1997)

[51] Yuan, C.: Some dierence schemes for the solution of the first boundary value problem for linear dierential equations with partial derivatives. Master's thesis, Moscow State University (1958)

i want morebooks!

Buy your books fast and straightforward online - at one of world's fastest growing online book stores! Environmentally sound due to Print-on-Demand technologies.

Buy your books online at

www.get-morebooks.com

Kaufen Sie Ihre Bücher schnell und unkompliziert online – auf einer der am schnellsten wachsenden Buchhandelsplattformen weltweit! Dank Print-On-Demand umwelt- und ressourcenschonend produziert.

Bücher schneller online kaufen

www.morebooks.de

VDM Verlagsservicegesellschaft mbH
Heinrich-Böcking-Str. 6-8   Telefon: +49 681 3720 174   info@vdm-vsg.de
D - 66121 Saarbrücken   Telefax: +49 681 3720 1749   www.vdm-vsg.de

Printed by Books on Demand GmbH, Norderstedt / Germany